Advances in Intelligent and Soft Computing 90

Editor-in-Chief: J. Kacprzyk

Advances in Intelligent and Soft Computing

Further volumes of this series can be found on our homepage: springer.com

Vol. 78. B.-Y. Cao, G.-J. Wang, S.-Z. Guo, and S.-L. Chen (Eds.)
Fuzzy Information and Engineering 2010
ISBN 978-3-642-14879-8

Vol. 79. A.P. de Leon F. de Carvalho, S. Rodríguez-González, J.F. De Paz Santana, and J.M. Corchado Rodríguez (Eds.)
Distributed Computing and Artificial Intelligence, 2010
ISBN 978-3-642-14882-8

Vol. 80. N.T. Nguyen, A. Zgrzywa, and A. Czyzewski (Eds.)
Advances in Multimedia and Network Information System Technologies, 2010
ISBN 978-3-642-14988-7

Vol. 81. J. Düh, H. Hufnagl, E. Juritsch, R. Pfliegl, H.-K. Schimany, and Hans Schönegger (Eds.)
Data and Mobility, 2010
ISBN 978-3-642-15502-4

Vol. 82. B.-Y. Cao, G.-J. Wang, S.-L. Chen, and S.-Z. Guo (Eds.)
Quantitative Logic and Soft Computing 2010
ISBN 978-3-642-15659-5

Vol. 83. J. Angeles, B. Boulet, J.J. Clark, J. Kovecses, and K. Siddiqi (Eds.)
Brain, Body and Machine, 2010
ISBN 978-3-642-16258-9

Vol. 84. Ryszard S. Choraś (Ed.)
Image Processing and Communications Challenges 2, 2010
ISBN 978-3-642-16294-7

Vol. 85. Á. Herrero, E. Corchado, C. Redondo, and Á. Alonso (Eds.)
Computational Intelligence in Security for Information Systems 2010
ISBN 978-3-642-16625-9

Vol. 86. E. Mugellini, P.S. Szczepaniak, M.C. Pettenati, and M. Sokhn (Eds.)
Advances in Intelligent Web Mastering – 3, 2011
ISBN 978-3-642-18028-6

Vol. 87. E. Corchado, V. Snášel, J. Sedano, A.E. Hassanien, J.L. Calvo, and D. Ślęzak (Eds.)
Soft Computing Models in Industrial and Environmental Applications, 6th International Workshop SOCO 2011
ISBN 978-3-642-19643-0

Vol. 88. Y. Demazeau, M. Pěchouček, J.M. Corchado, and J.B. Pérez (Eds.)
Advances on Practical Applications of Agents and Multiagent Systems, 2011
ISBN 978-3-642-19874-8

Vol. 89. J.B. Pérez, J.M. Corchado, M.N. Moreno, V. Julián, P. Mathieu, J. Canada-Bago, A. Ortega, and A.F. Caballero (Eds.)
Highlights in Practical Applications of Agents and Multiagent Systems, 2011
ISBN 978-3-642-19916-5

Vol. 90. J.M. Corchado, J.B. Pérez, K. Hallenborg, P. Golinska, and R. Corchuelo (Eds.)
Trends in Practical Applications of Agents and Multiagent Systems, 2011
ISBN 978-3-642-19930-1

Juan M. Corchado, Javier Bajo Pérez,
Kasper Hallenborg, Paulina Golinska, and
Rafael Corchuelo (Eds.)

Trends in Practical Applications of Agents and Multiagent Systems

9th International Conference on Practical Applications of Agents and Multiagent Systems

Springer

Editors

Prof. Juan M. Corchado
Universidad de Salamanca
Departamento de Informática y Automática
Facultad de Ciencias
Plaza de la Merced S/N
37008 Salamanca
Spain
E-mail: corchado@usal.es

Prof. Javier Bajo Pérez
Universidad Pontificia de Salamanca
Escuela Universitaria de Informática
Compañía 5
37002 Salamanca
Spain

Prof. Kasper Hallenborg
University of Southern Denmark
The Maersk Mc-Kinney Moller Institute
Campusvej 55
DK-5230 Odense M
Denmark

Prof. Paulina Golinska
Poznan University of Technology
Insitute of Manangement Engineering
Strzelecka 11
60-965 Poznan
Poland
E-mail: paulina.golinska@put

Prof. Rafael Corchuelo
ETSI Informática
Avda. Reina Mercedes, s/n
Sevilla, 41012
Spain
E-mail: corchu@us.es

ISBN 978-3-642-19930-1 e-ISBN 978-3-642-19931-8

DOI 10.1007/978-3-642-19931-8

Advances in Intelligent and Soft Computing ISSN 1867-5662

Library of Congress Control Number: 2011923220

Typeset & Cover Design: Scientific Publishing Services Pvt. Ltd., Chennai, India.

Printed on acid-free paper
5 4 3 2 1 0
springer.com

Preface

PAAMS'11 Workshops complement the regular program and the special sessions with new or emerging trends of particular interest connected to multi-agent systems.

PAAMS, the International Conference on Practical Applications of Agents and Multi-Agent Systems is an evolution of the International Workshop on Practical Applications of Agents and Multi-Agent Systems. PAAMS is an international yearly tribune to present, to discuss, and to disseminate the latest developments and the most important outcomes related to real-world applications. It provides a unique opportunity to bring multi-disciplinary experts, academics and practitioners together to exchange their experience in the development of Agents and Multi-Agent Systems.

This volume presents the papers that have been accepted for the 2011 in the workshops: Workshop on Agents for Ambient Assisted Living, Workshop on Agent-Based Solutions for Manufacturing and Supply Chain, Workshop on Agents and Multi-agent systems for Enterprise Integration.

We would like to thank all the contributing authors, as well as the members of the Program Committees of the workshops and the Organizing Committee for their hard and highly valuable work. Their work has helped to contribute to the success of the PAAMS'11 event. Thanks for your help, PAAMS'11 wouldn't exist without your contribution.

Juan Manuel Corchado
Javier Bajo
PAAMS'11 Organizing Co-chairs

Acknowledgements: This conference has been supported by Junta de Castilla y León (Spain).

Organization

Workshops

W1 – Workshop on Agents for Ambient Assisted Living
W2 – Workshop on Agent-Based Solutions for Manufacturing and Supply Chain
W3 – Workshop on Agents and Multi-agent systems for Enterprise Integration

Workshop on Agents for Ambient Assisted Living Committee

Kasper Hallenborg	University of Southern Denmark (Denmark)
Sylvain Giroux	Université de Sherbrooke (Canada)
Bruno Bouchard	Université Paris-Dauphine (France)
Fabrice Peyrard	Toulouse II University (France)
Nikolaos Spanoudakis	Technical University of Crete (Greece)
Olivier Boissier	ENS Mines de Saint-Etienne (France)
Patrice Roy	Unversity of Sherbrooke (Canada)
Pierre Busnel	Unversity of Sherbrooke (Canada)
Valerie Camps	University of Toulouse (France)

Workshop on Agent-Based Solutions for Manufacturing and Supply Chain Committee

Mathieu Philippe (Co-chairman)	Lille1 University (France)
Florian Hauser	University of Innsbruck (Austria)
Olivier Brandouy	University Paris 1 Panthéon Sorbonne (France)
Silvano Cincotti	University of Genoa (Italy)

Workshop on Agents and Multi-agent systems for Enterprise Integration (ZOCO 2011) Committee

Organizing Committee

José L. Álvarez	Universidad de Huelva (Spain)
Manuel Álvarez	Universidad de la Coruña (Spain)

Scientific Committee

Vijay Sugumaran	Oakland University (USA)
Tzouramanis Theodoros	University of Aegean (Greece)
María E. Vidal	University of Maryland (USA)
Graham Winstanley	University of Brighton (UK)
Gracja Wydmuch	Wroclaw University of Economics (Poland)
Sadok Ben Yahia	Faculté des Sciences de Tunis (Tunisia)
Hongji Yang	De Montfort University (UK)

Organizing Committee

Juan M. Corchado (Chairman)	University of Salamanca (Spain)
Javier Bajo (Co-Chairman)	Pontifical University of Salamanca (Spain)
Juan F. De Paz	University of Salamanca(Spain)
Sara Rodríguez	University of Salamanca (Spain)
Dante I. Tapia	University of Salamanca (Spain)
Emilio Corchado	University of Salamanca (Spain)
Fernando de la Prieta Pintado	University of Salamanca (Spain)
Davinia Carolina Zato Domínguez	University of Salamanca (Spain)

Contents

Workshop on Agents for Ambient Assisted Living

Improving Context-Awareness in a Healthcare Multi-Agent System 1
Ricardo S. Alonso, Dante I. Tapia, Óscar García, David Sancho, Miguel Sánchez

Elder Care Alert Management - Decision Support by a Logistic Regression Model 9
Alexandra Seco, Miguel Felgueiras, Florentino Fdez-Riverola, António Pereira

A New Adaptive Algorithm for Detecting Falls through Mobile Devices 17
Miguel Sánchez, Patricia Martín, Laura Álvarez, Vidal Alonso, Carolina Zato, Alberto Pedrero, Javier Bajo

Agent-Supported Assessment for Adaptive and Personalized Ambient Assisted Living 25
Helena Lindgren, Dipak Surie, Ingeborg Nilsson

A Multi-Agent System for Remote Eldercare 33
Boštjan Kaluža, Erik Dovgan, Violeta Mirchevska, Božidara Cvetković, Mitja Luštrek, Matjaž Gams

iRoute: Cognitive Support for Independent Living Using BDI Agent Deliberation 41
Shabbir Hossain, Kasper Hallenborg, Yves Demazeau

Dynamic Distribution of Tasks in Health-Care Scenarios 51
Carolina Zato, Ana de Luis, Juan F. De Paz

Workshop on Agent-Based Solutions for Manufacturing and Supply Chain (AMSC'11)

Automating the Development of Agent-Based Warehouse Control Systems 59
Jacques Verriet, Bruno van Wijngaarden, Eef van Heusden, Roelof Hamberg

Manufacturing Material Flow Analysis Based on Agent and Movable Resource Concept 67
Pawel Pawlewski

How to Successfully Combine Case Based Reasoning and Multi-Agent Systems for Supply Chain Improvement 75
Paul-Eric Dossou, Philip Mitchell, Pawel Pawlewski

Market Strategy Choices Made by Company Using Reinforcement Learning 83
Dawid Chodura, Paweł Dominik, Jarosław Koźlak

Myopic Behaviour in Holonic Multiagent Systems for Distributed Control of FMS 91
Emmanuel Adam, Gabriel Zambrano, Cyrille Pach, Thierry Berger, Damien Trentesaux

Agent-Based System for Planning and Coordination of Intermodal Transport 99
Paulina Golinska, Marcin Hajdul

Workshop on Agents and Multi-Agent Systems for Enterprise Integration (ZOCO'11)

Boosting Bitext Compression 109
Joaquín Adiego, Miguel A. Martínez-Prieto, Javier E. Hoyos-Torío, Felipe Sánchez-Martínez

Web Document Duplicate Detection Using Fuzzy Hashing ... 117
Carlos G. Figuerola, Raquel Gómez Díaz, José L. Alonso Berrocal, Angel F. Zazo Rodríguez

An Implementation of a Trading Service for Building Open and Interoperable DT Component Applications 127
Carmelo Maturana, Antonio Jesús Fernández-García, Luis Iribarne

Towards Semi-automatic Construction of Data Mashups 137
Javier López, Fernando Bellas, Alberto Pan, Paula Montoto

Adding Semantics to the Discovery of Web Services in Life Sciences 145
María Pérez, Ismael Sanz, Rafael Berlanga, María José Aramburu

Bringing Statistical Methodologies for Enterprise Integration of Conversational Agents 153
David Griol, José Manuel Molina

Mining Web Pages Using Features of Rendering HTML Elements in the Web Browser 161
F.J. Fernández, José L. Álvarez, Pedro J. Abad, Patricia Jiménez

An Experiment on Using Datamining Techniques to Extract Information from the Web 169
Gretel Fernández, Hassan A. Sleiman

BRENHET2, A MAS to Facilitate the Reutilization of LOs through Federated Search 177
F. De la Prieta, A. Gil, S. Rodríguez, B. Martín

Author Index 185

Improving Context-Awareness in a Healthcare Multi-Agent System

Ricardo S. Alonso, Dante I. Tapia, Óscar García,
David Sancho, and Miguel Sánchez

Abstract. Context-aware technologies allow Ambient Assisted Living systems and applications to automatically obtain information from users and their environment in a distributed and ubiquitous way. One of the most important technologies used to provide context-awareness to a system is Wireless Sensor Networks. This paper describes last improvements made on ALZ-MAS, an Ambient Intelligence based multi-agent system aimed at enhancing the assistance and healthcare for Alzheimer patients. In this sense, a new ZigBee platform is used to improve ALZ-MAS. This platform provides the system with new telemonitoring and locating engines that facilitate the integration of context-awareness into it.

Keywords: Multi-Agent Systems, Ambient Intelligence, Context-Awareness, Wireless Sensor Networks, Real-Time Locating Systems, e-Healthcare.

1 Introduction

Dependence is a permanent situation in which a person needs important assistance from others in order to perform basic daily life activities such as essential mobility, object and people recognition, as well as domestic tasks [2]. There is an ever growing need to supply constant care and support to the disabled and elderly, and the drive to find more effective ways of providing such care has become a major challenge for the scientific community. In this sense, Ambient Intelligence (AmI) is an emerging multidisciplinary area based on ubiquitous computing, which

---

Ricardo S. Alonso · Dante I. Tapia · Óscar García · David Sancho
Nebusens, S.L. Scientific Park of the University of Salamanca, Calle Adaja, s/n, 37185, Villamayor de la Armuña, Salamanca, Spain.
e-mail: {ricardo.alonso,dante.tapia,oscar.garcia,david.sancho}@nebusens.com

Miguel Sánchez
Pontifical University of Salamanca, Calle Compañía, 5, 37002, Salamanca, Spain

J.M. Corchado et al. (Eds.): Trends in PAAMS, AISC 90, pp. 1–8.
springerlink.com

influences the design of protocols, communications, systems and devices, proposing new ways of interaction between people and technology, thus adapting them to the needs of individuals and their environment [13].

One key aspect in any AmI-based system is the use of context-aware technologies. The context is defined as any information used to characterize the situation of an entity, which can be a person, a place or an object [3]. This information is important for defining the interaction between users and the technology that surrounds them. However, it is not enough to gather information about the context, but that information must be processed by self-adaptable and dynamic mechanisms and methods that can react independently of each particular situation that arises. In this sense, Multi-Agent Systems (MAS) [14] comprise one of the areas that can contribute expanding the possibilities of Ambient Assisted Living [8]. Furthermore, most of the context information can be collected by distributed sensors throughout the environment and even by the users themselves. Wireless Sensor Networks (WSN), such as ZigBee [15], are more flexible and require less infrastructural support than wired sensor networks [9], avoiding the need of wiring homes or hospitals and decreasing the costs and drawbacks of the setup phase.

This paper describes the new changes introduced into ALZ-MAS [1], an Ambient Intelligence based multi-agent system aimed at enhancing the assistance and healthcare for Alzheimer patients in geriatric residences. An essential aspect in this work is the use of Wireless Sensor Networks to provide the agents with automatic and real-time information of the environment and allow them to react upon it. Therefore, these new changes include the integration of n-Core [7], an innovative wireless sensor platform, to improve the context-awareness of the system with better and new telemonitoring and locating techniques.

The problem description that motivated the development of ALZ-MAS is presented in the next section. Then, it is described the basic components of ALZ-MAS and the new changes introduced in the system to provide its agents with improved context-aware capabilities. In addition, it is explained some experiments made to evaluate the performance of the system applied to a real scenario, as well as the obtained results, comparing the previous version of ALZ-MAS and the new release presented in this paper. Finally, conclusions are depicted.

2 Problem Description

Ambient Intelligence offers a great potential to improve quality of life and simplify the use of technology by offering a wider range of personalized services and providing users with easier and more efficient ways to communicate and interact with other people and systems [13]. Among the general population, those most likely to benefit from the development of these systems are the elderly and dependent people, whose daily lives, with particular regard to healthcare, will be most enhanced [11]. Dependent people can suffer from degenerative diseases, dementia, or loss of cognitive ability [2]. However, the development of systems that clearly fulfill the needs of AmI is difficult and not always satisfactory. It requires a joint development of models, techniques and technologies based on services.

As mentioned before, one of the key aspects for the construction of AmI-based systems is obtaining information from the context through sensor networks. The context includes information about people and their environment. The information may consist of many different parameters such as location, building status (*e.g.*, temperature), vital signs (*e.g.*, heart rhythm), etc.

Sensor networks are presented as an optimal technology for obtaining contextual information. Wired sensor networks, such as X10, LonWorks or KNX, are not as flexible as wireless sensors networks, such as ZigBee, and require more infrastructural support [9]. The ZigBee standard [15] operates in the 868/915MHz and 2.4GHz unlicensed bands. ZigBee is designed to work with low-power nodes and allows up to 65,534 nodes to be connected in a star, tree or mesh topology network. There are several telemonitoring healthcare developments based on Wireless Sensor Networks [6, 12]. However, these developments do not take into account their integration with other systems or architectures and are difficult to be adapted to other different scenarios.

Furthermore, the information obtained must be managed by intelligent and self-adaptable technologies to provide an adequate interaction between the users and their environment. Agents and Multi-Agent Systems are one of these technologies. Agents have a set of characteristics, such as autonomy, reasoning, reactivity, social abilities, pro-activity, mobility or organization which allow them to cover several needs for Ambient Assisted Living environments, as they possess the capability of adapting themselves to the users and environmental characteristics [5]. Multi-agent systems and architectures based on intelligent devices have recently been used to provide assistance to elderly and dependent people [8]. Although these applications expand the possibilities and stimulate research efforts to enhance the assistance and healthcare provided to elderly and dependent people, none of them integrate intelligent agents, distributed and dynamic applications and services approach or the use of reasoning and planning mechanisms.

3 ALZ-MAS: An Alzheimer Healthcare Multi-Agent System

ALZ-MAS (*ALZheimer Multi-Agent System*) [1, 10] is a distributed multi-agent system designed upon Ambient Intelligence and aimed at enhancing the assistance and healthcare for Alzheimer patients living in geriatric residences. The main functionalities in the system include reasoning and planning mechanisms [4] that are embedded into deliberative BDI (Belief-Desire-Intention) agents, and the use of several context-aware technologies to acquire information from users and their environment. The current version of the system, ALZ-MAS 2.0, makes use of FUSION@ (*Flexible User and ServIces Oriented multiageNt Architecture*), an architecture that combines a SOA approach with intelligent agents for building highly dynamic Ambient Intelligence based systems [10].

As can be seen in Figure 1, ALZ-MAS structure has five different deliberative agents based on the BDI model (BDI Agents), each one with specific roles and capabilities: *User Agent*, *SuperUser Agent*, *ScheduleUser Agent*, *Admin Agent* and *Devices Agent*. In ALZ-MAS, each agent integrates its own functionalities into their structure. If an agent needs to perform a task which involves another agent, it must communicate with that agent to request it. So, if the agent is disengaged, all its functionalities will be unavailable to the rest of agents. Another important issue is that the reasoning (CBR) and planning (CBP) mechanisms are integrated into the agents. The main problem with ALZ-MAS 1.0 was the efficiency of the computational planning processes as well as the integration with other existing architectures. This way, the main components of ALZ-MAS 2.0 were modeled as distributed and independent services by means of FUSION@, releasing the agents from high demanding computational processes. FUSION@ proposes a new perspective, where Multi-Agent Systems and SOA-based services are integrated to provide ubiquitous computation, ubiquitous communication and intelligent interfaces facilities. As can be seen in Figure 1, the FUSION@ architecture defines four basic blocks: *Applications*, *Services*, *Agents Platform* and *Communication Protocol* [10]. Besides, there are seven pre-defined agents that provide the basic functionalities of the architecture: *CommApp*, *CommServ*, *Directory*, *Supervisor*, *Security*, *Admin* and *Interface*. *Interface Agents* were designed to be embedded in users' applications and are simple enough to allow them to be executed on mobile devices, such as cell phones or PDAs.

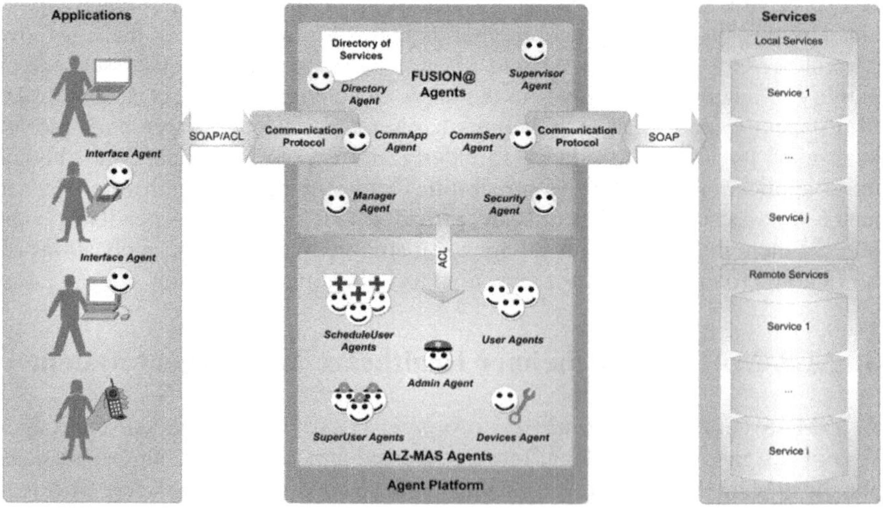

Fig. 1 Schema of the ALZ-MAS 2.0 system

Agents in ALZ-MAS 2.0 [10] collaborate with context-aware agents that employ Radio Frequency IDentification (RFID) [1], Wireless Sensor Networks and automation devices to provide automatic and real-time information about the environment, and allow users to interact with their surroundings, controlling and managing physical services (*i.e.*, heating, lights, switches, etc.). All the information provided is processed by the agents, specially the *Devices Agent* which is a BDI agent that runs on a workstation. All hardware is someway integrated with agents, providing automatic and real-time information about the environment that is processed by the agents to automate tasks and manage multiple services.

The configuration used in ALZ-MAS 2.0 consisted of an RFID transponder mounted on a bracelet worn on the users' wrist or ankle, several RFID readers installed over protected zones with an adjustable capture range up to 2 meters, and a central workstation where all the information is processed and stored. These RFID readers were installed over the doors accessing to restricted zones so that the system could detect when a user was trying to enter a forbidden area according to the user's permissions. Nevertheless, this configuration does not allow knowing the exact position of each user in real-time. Moreover, this approach requires gathering information from two distinct sensor networks: an RFID network for detecting unauthorized accesses and other ZigBee network to obtain information of the environment from different sensors, such as temperature sensors, light sensors, as well as smoke and gas detectors. Thus, the new ALZ-MAS 2.1 has been improved with the integration of an innovative ZigBee platform into the system to provide it with enhanced telemonitoring and new locating features.

3.1 Improving the Context-Aware Technologies Used in ALZ-MAS 2.0

As mentioned before, ALZ-MAS 2.0 used a combination of RFID and ZigBee technologies to obtain context information. This way, while RFID tags and readers was used to allow the system to detect unauthorized accesses to restricted areas, a ZigBee network was used to provide the system with sensing capacities. Nevertheless, the new ALZ-MAS 2.1 uses just ZigBee devices to obtain both users' location and other context information.

Figure 2 (left) shows the ZigBee devices used in ALZ-MAS 2.1 to control heating, lights, door locks, alarms, etc. These devices, called n-Core Sirius-A, have both 2.4GHz and 868/915MHz versions and have several communication ports (GPIO, ADC, I2C and UART through USB or DB-9 RS-232) to connect to distinct devices, including a wide range of sensors and actuators [7]. This way, it is deployed a mesh of these devices to control all these services. n-Core Sirius-A devices are part of the n-Core platform (http://www.n-core.info), which offers a complete API (Application Programming Interface) to access all its functionalities, including telemonitoring and locating engines. This API can be easily accessed as a Dynamic Link Library (DLL) from different languages and platforms, such as C/C++, .Net, Java or Python. Thus, it can be used different locating techniques using Sirius A devices as readers and Sirius B devices, showed in Fig. 2 (right), as tags carried by patients and medical personnel. Sirius B devices are small enough to be carried by a patient, a caregiver or even an object, and provide a battery life of up to two years.

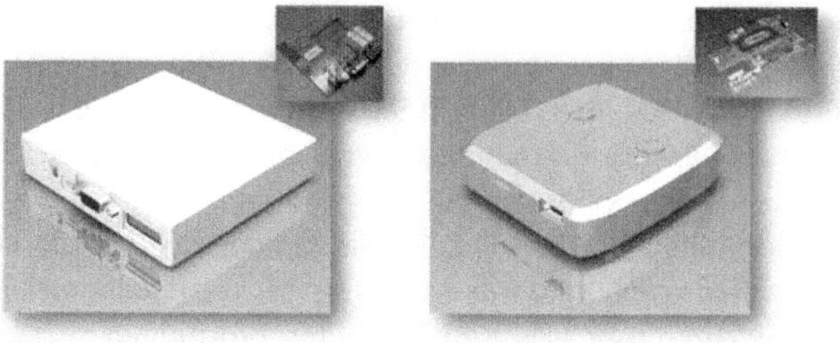

Fig. 2 n-Core Sirius A (left) and Sirius B (right) devices used in ALZ-MAS 2.1

In ALZ-MAS 2.0, the system could only know in which room each user was, because location was based on knowing which door (*i.e.*, RFID antenna) a user was passed through. In ALZ-MAS 2.1 the location of users is given as coordinated points obtained from the locating techniques provided by the n-Core API. All information obtained by means of these technologies is processed by the agents. Nonetheless, thanks to the architecture of the system, it was not necessary to introduce any change into the agents' reasoning mechanisms. Depending on the system requirements, several interfaces can be executed. The interfaces show basic information about nurses and patients (name, tasks that must be accomplished, location inside the residence, etc.) and the building (specific room temperature, lights status, etc.). Figure 3 shows one of the new user interfaces of ALZ-MAS 2.1., where it is shown the positions of the users given by the new locating engine.

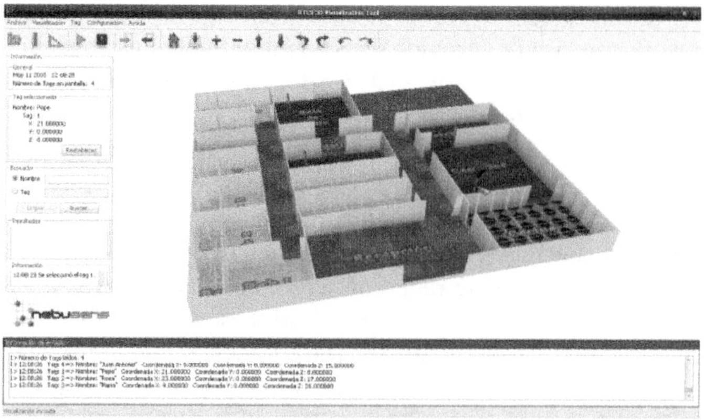

Fig. 3 Graphical User Interface of the ALZ-MAS 2.1

After the inclusion of n-Core into ALZ-MAS, it has been done new experiments to test the improvements made in the context-awareness of ALZ-MAS. This way, the new ALZ-MAS 2.1 prototype using new n-Core ZigBee devices and API [7] for both locating and telemonitoring tasks was implemented in the same center [1] and compared with previous ALZ-MAS 2.0, using both RFID and ZigBee devices. One of these experiments consisted of testing the performance of

ALZ-MAS when detecting users accessing to restricted areas according to their permissions. ALZ-MAS monitors and locate the patients and guarantees that each one of them is in the right place, and secondly, only authorized personnel can gain access to the residence protected areas. Figure 4 shows the number of accesses to restricted zones detected before the implementation of ALZ-MAS 2.1 with the previous release of ALZ-MAS 2.0 during a 30-day period. With the new release of ALZ-MAS 2.1 using new ZigBee devices and more accurate locating techniques, there is an improvement regarding the previous ALZ-MAS version, increasing from 3 to 8 daily detections. Furthermore, thanks to the new innovative locating techniques, the number of daily false positives has been reduced from 7 to 0.87.

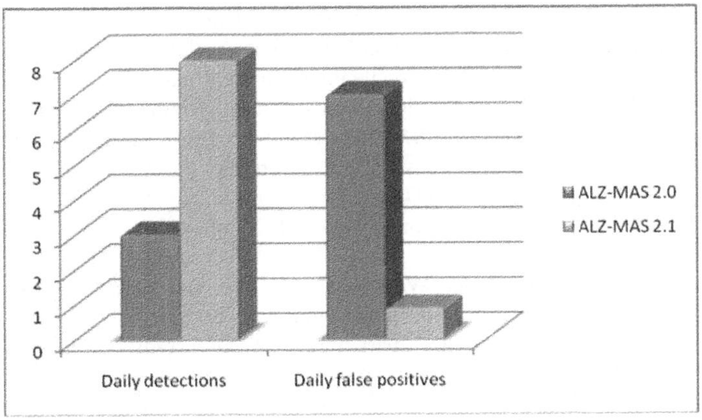

Fig. 4 Comparison of the daily number of detected accesses to restricted areas and false positives between ALZ-MAS 2.0 and ALZ-MAS 2.1

4 Conclusions

ALZ-MAS can improve the security and healthcare efficiency through monitoring medical staff's work and patients' activities. Context-aware technologies provide ALZ-MAS with valuable information about the environment, contributing to a ubiquitous and non-invasive interaction among users, the system and the environment. Thanks to the n-Core platform, the new ALZ-MAS 2.1 improves significantly the context information in respect of ALZ-MAS 2.0. Qualitatively, the location information is provided as coordinated positions, instead of just which room each user is. Quantitatively, security has been clearly improved. Furthermore, infrastructure costs have been reduced in ALZ-MAS 2.1 in relation to ALZ-MAS 2.0. Although each RFID transponder is cheaper than a ZigBee device used as tag, RFID antennas are quite more expensive than ZigBee devices used as readers. Even more, to obtain the same level of precision in the location information than the provided by the ZigBee platform, it would be necessary many more expensive RFID readers.

Acknowledgments. This work has been supported by Spanish Ministry of Science and Innovation Project Ref. TRA2009_0096.

References

1. Corchado, J.M., Bajo, J., Paz, Y.D., Tapia, D.I.: Intelligent environment for monitoring Alzheimer patients, agent technology for health care. Decis. Support Syst. 44(2), 382–396 (2008)
2. Costa-Font, J., Patxot, C.: The design of the long-term care system in Spain: Policy and financial constraints. Social Policy and Society 4(1), 11–20 (2005)
3. Dey, A.K., Abowd, G.D.: Towards a Better Understanding of Context and Context-Awareness. In: Proceedings of the CHI 2000 Workshop on The What, Who, Where, When, and How of Context-Awareness. The Hague, Netherlands (2000)
4. Glez-Bedia, M., Corchado, J.M.: A planning Strategy based on Variational Calculus for Deliberative Agents. Computing and Information Systems Journal 10, 2–14 (2002)
5. Jayaputera, G.T., Zaslavsky, A.B., Loke, S.W.: Enabling run-time composition and support for heterogeneous pervasive multi-agent systems. Journal of Systems and Software 80(12), 2039–2062 (2007)
6. Jurik, A.D., Weaver, A.C.: Remote Medical Monitoring. Computer 41(4), 96–99 (2008)
7. n-Core: A Faster and Easier Way to Create Wireless Sensor Networks (2010), `http://www.n-core.info` (retrieved January 9, 2011)
8. Pecora, F., Cesta, A.: Dcop for smart homes: A case study. Computational Intelligence 23(4), 395–419 (2007)
9. Sarangapani, J.: Wireless Ad hoc and Sensor Networks: Protocols, Performance and Control. CRC, Boca Raton (2007)
10. Tapia, D.I., Bajo, J., Corchado, J.M.: Distributing Functionalities in a SOA-Based Multi-agent Architecture. In: 7th International Conference on Practical Applications of Agents and Multi-Agent Systems (PAAMS 2009), pp. 20–29 (2009)
11. Van Woerden, K.: Mainstream Developments in ICT: Why are They Important for Assistive Technology? Technology and Disability 18(1), 15–18 (2006)
12. Varshney, U.: Improving Wireless Health Monitoring Using Incentive-Based Router Cooperation. Computer 41(5), 56–62 (2008)
13. Weber, W., Rabaey, J.M., Aarts, E.: Ambient Intelligence. Springer-Verlag New York, Inc., Heidelberg (2005)
14. Wooldridge, M.: An Introduction to MultiAgent Systems. John Wiley & Sons, Chichester (2002)
15. ZigBee Alliance, ZigBee Specification Document 053474r13. ZigBee Standards Organization (2006)

Elder Care Alert Management - Decision Support by a Logistic Regression Model

Alexandra Seco, Miguel Felgueiras, Florentino Fdez-Riverola, and António Pereira

Abstract. The Elder Care project implies the use of a body sensor and a location device by each elderly. This architecture will originate a very large data set that should be treated by statistical methods. We purpose a logistic model approach, that will be used during the prototype implementation phase and after that in the comercial phase. In the prototype implementation phase (already running), an elders sample is going to be analyzed to eliminate some redundant or useless covariates, allowing a spare in the final equipment complexity and price. Besides, logistic model has an easy odds ratio interpretation, and so we will be able to determine elders problems risk factors. After the prototype implementation phase, we hope to have a better knowledge of elders risk factors and important covariates that must be considered. Again, the logistic model may be applied to correctly decide whenever an alarm signal should be sent for some monitorized elderly, preventing sensibility and specificity issues. The presented decision model will be used by a multi-agent system, in order to make reliable decisions using fewer resources.

Keywords: health monitoring, logistic regression, multi-agent systems, odds ratio.

Alexandra Seco · António Pereira
School of Technology and Management,
Computer Science and Communications Research Centre,
Polytechnic Institute of Leiria, P-2411-901, Leiria, Portugal
e-mail: alexandra.seco@ipleiria.pt, apereira@ipleiria.pt

Miguel Felgueiras
CEAUL Lisbon and School of Technology and Management,
Computer Science and Communications Research Centre,
Polytechnic Institute of Leiria, P-2411-901, Leiria, Portugal
e-mail: mfelg@ipleiria.pt

Florentino Fdez-Riverola
ESEI: Escuela Superior de Ingeniera Informática, University of Vigo,
Edificio Politcnico, Campus Universitario As Lagoas s/n, 32004, Ourense, Spain
e-mail: riverola@uvigo.es

J.M. Corchado et al. (Eds.): Trends in PAAMS, AISC 90, pp. 9–16.
springerlink.com

1 Introduction

It is a fact that global life expectancy is rising and the number of elderly people is increasing. Elderly people show a strong determination to live at home as long as possible, and have their own independence. But, on the other side, they also demonstrate a fear of feeling unsafe and lonely, and have health issues, and this becomes a problem to the families. However, with the lack of time of families and life style, elders care may frequently became a great problem to all.

Wireless communications made remote monitoring of people possible, allowing us to have 24 hours of surveillance with low costs. In this context, Elder Care architecture was proposed by Marcelino *et al* (2008). The Elder Care solution has two primary goals: monitoring vital signs, sending alerts to family and specialized help and provide a social network in order to wrap all elder to avoid social isolation (Marcelino *et al*, 2010). They propose an architecture to monitor vital signs giving the proper response in abnormality cases. The local monitoring will be installed in elderly people's home. Each user will have a set of wireless body sensor that will allow continuous vital parameter monitoring such as blood pressure and fall detection. Some routines, like the time spent at home and outside patch will also be monitored. This information will be sent to a control center that will be responsible for setting the proper alerts in abnormal situations. If the elderly is in their own residence, the information will be sent through a communication network. When the elderly is outside residence, the information will be sent through their cell phone (Marcelino and Pereira, 2009).

On the other hand, and as stated by Bajo *et al* (2010), multi-agent systems are considered a key issue for the success of agent technology in real world scenarios. In fact, as already noticed by Augusto and McCullagh (2007), multi-agent systems can facilitate the development of pervasive home care environments since they require effective communication as well as distributed problem solving.

Having these requirements in mind, the multi-agent systems can be used to improve the Elder Care architecture by optimizing the system resources. However, some problems still exist. For example: how to identify abnormal situations? How to make a frontier from signs that are normal situations and relegate abnormal situations? For instance, if the elderly sits strongly, how to identify if this is normal or a fall that should be reported to family or specialized help? The problems of reducing false positives is an important one, for the reasons pointed above and furthermore improves efficiency in the network as it decreases energy consume.

Since the response is to send an alert or do not send a alert, we propose a logit statistical model to make decisions based on the information. Considering Y as the response variable, it will only take values 0 or 1, where $Y = 1$ represents the situation when an alarm signal should be sent to the control center and $Y = 0$ the situation when such alarm is unnecessary. As the logistic model is a generalized linear model, an extended and versatile class of models, we will begin to present their theoretical scope.

2 Generalized Linear Models

Generalized linear models (GLM) were previously presented by Nelder and Wedderburn (1972) and later developed widely by McCullagh and Nelder in 1983. GLM include as special cases linear regression, logit and probit models for quantal responses, loglinear models for counts and some more commonly used statistical models. These authors, instead of noticing differences between the referred models, shown that they share a number of properties, such as linearity and parameters estimation, and distributions belonging to the exponential family, so that they can be considered as a single class named GLM.

We will present this class and later focus on the logistic (or logit) models, as they are the chosen one for the present elders problem.

2.1 *Definition*

Suppose that Y_i, $i = 1,...,N$ are N random independent variables and $\mu_i = E(Y_i)$, $i = 1,...,N$. An GLM has three parts:

1. The random component. Each Y_i, $i = 1,...,N$, has distribution f in the exponential family taking the form

$$f(y_i|\theta_i,\phi) = \exp\left\{\frac{y_i\theta_i - b(\theta_i)}{a_i(\phi)} + c(y_i,\phi)\right\},$$

   for some specific functions $a(.), b(.)$, $c(.,.)$ with canonical parameter θ_i. ϕ is a scale parameters that does not depends on i, usually called dispersion parameter, and $a_i(\phi)$ is usually written as ϕ/w_i, where w_i is a prior weight.
2. The systematic component. A linear predictor η_i relating the covariates $x_{i1},...,x_{ip}$,

$$\eta_i = \sum_{j=1}^{p} x_{ij}\beta_j, \quad i = 1,...,N,$$

   where β_j, $j = 1,...,p$ are unknown parameters to be estimated.
3. The link function g, monotonic and differentiable, relating the linear predictor with μ_i,

$$\eta_i = g(\mu_i),\ i = 1,...,N. \tag{1}$$

As stated before, this models generalize the linear regression model and logit models where $\eta_i = \log(\frac{\mu_i}{1-\mu_i})$.

2.2 *Fitting Procedure - Estimating β and Models Selection*

Maximum-likelihood estimates will be used and can be obtained by iterative weighted least squares. From the $Y_1,...,Y_N$ log-likelihood function,

$$l = \sum_{i=1}^{N} \left[\frac{y_i \theta_i - m(\theta_i)}{a_i(\phi)} + c(y_i, \phi) \right].$$

By the chain rule (Seco, 2003)

$$\frac{\partial l}{\partial \beta_j} = \sum_{i=1}^{N} \frac{(y_i - \mu_i)x_{ij}}{var(Y_i)} \left(\frac{d\mu_i}{d\eta_i} \right), \; j = 1, ..., p.$$

The equations $\frac{\partial l}{\partial \beta_j} = 0$, $j = 1, ..., p$, are non-linear in $\beta_1, ..., \beta_p$, but can be solved by Newton-Raphson method.

In practice, the iterative method has the following steps:

1. Let $t = 0$. Consider an initial estimate $\hat{\mu}_i^{(0)}$, $i = 1, ..., N$.
2. Compute

$$\hat{z}_i^{(t)} = \hat{\eta}_i^{(t)} + (y_i - \hat{\mu}_i^{(t)}) \left(\frac{d\eta_i}{d\mu_i} \right)^{(t)},$$

and

$$\hat{w}_{ii}^{(t)} = \frac{1}{a_i(\phi) V(\hat{\mu}_i^{(t)})} \left[\left(\frac{d\mu_i}{d\eta_i} \right)^2 \right]^{(t)}, \; i = 1, ..., N.$$

3. Compute

$$\hat{\beta}^{(t+1)} = (\mathbf{X}^T \hat{\mathbf{W}}^{(t)} \mathbf{X})^{-1} \mathbf{X}^T \hat{\mathbf{W}}^{(t)} \hat{\mathbf{z}}^{(t)},$$
$$\hat{\eta}^{(t+1)} = \mathbf{X} \hat{\beta}^{(t+1)}, \quad \hat{\mu}^{(t+1)} = g^{-1}(\hat{\eta}^{(t+1)}).$$

4. Compare $\hat{\beta}^{(t-1)}$ with $\hat{\beta}^{(t)}$. If they are sufficiently closed, stop. If not, let $t = t + 1$ and go to step 2.

The scaled deviance, as defined by McCullagh and Nelder (1989), compares the log-likelihood of the current model with the saturated models which has N parameters:

$$D^*(\mathbf{y}|\hat{\mu}) = 2 \left[l(\mathbf{y}, \phi | \mathbf{y}) - l(\hat{\mu}, \phi | \mathbf{y}) \right].$$

Model selection is then based on asymptotic distribution of D^* (Seco, 2003), choosing the simplest model that fits well enough.

3 The Logistic Model

The logistic model is a GLM that presents a linear predictor (logit) defined as

$$\ln \left(\frac{\pi}{1 - \pi} \right) = \beta_0 + \beta_1 x_1 + ... + \beta_p x_p$$

where

$$\pi = P(Y = 1).$$

For the Elder Care context, the Y variable only takes values 0 or 1, because we are interesting in detecting whenever the elderly has a problem. Therefore, $Y = 1$ represents the situation when an alarm signal should be sent to the control center and $Y = 0$ the situation when such alarm is unnecessary.

Each variable X_i, $i = 1, ..., p$, represent a vital parameter monitorized by a body sensor or a location device. When some combination of $(X_1, ...X_n)$ suggests that the user is having a problem (lost or hurt), the estimated value of π should be greater than 0.5, and therefore the alarm is trigged.

The β_i parameters are connected with the importance of the respective X_i variables, and represent the logit change produced by an unitary increase in X_i. assuming for simplicity purposes that X_i has a linear effect and that no interactions between covariate X_i variables are relevant. Under this conditions,

$$[\beta_0 + \beta_1 x_1 + ... + \beta_i (x_i + 1) + .. + \beta_p x_p] - [\beta_0 + \beta_1 x_1 + ... + \beta_i x_i + .. + \beta_p x_p] = \beta_i$$

as stated.

4 Odds and Odds Ratio

4.1 Odds

Let

$$\pi_i = P(Y = 1|\mathbf{x}_i)$$

where

$$\mathbf{x}_i^T = (x_{i1}, ..., x_{ip}).$$

An odd is descrived as

$$O_i = \frac{\pi_i}{1 - \pi_i}.$$

So, for each elderly, the possibility of a problem to occur is simply the quotient between the probabilities of success and insuccess. Under this circunstancies, it is straightforward to note that if $\pi_i > 0.5$ then $O > 1$ and if $\pi_i < 0.5$ then $0 < O < 1$. For instance, if $\pi_i = 0.2$ then

$$O_i = \frac{0.2}{0.8} = \frac{1}{4},$$

and we might say the odds of an elderly problem occur are 1 to 4.

4.2 Odds Ratio

Probably the more important feature of the logistic model is the simplicity of the odds ratio calculation and interpretation that is achieved under this model. An odds ratio is defined as the quotient between to possibilities, that is

$$OR = \frac{O_1}{O_2} = \frac{\frac{\pi_1}{1-\pi_1}}{\frac{\pi_2}{1-\pi_2}}.$$

Clearly, if $OR > 1$ than the first data set has higher odds than the second, and if $OR < 1$ we have the reverse situation.

For interpretation purposes, let us considerer a simple model

$$\ln\left(\frac{\pi}{1-\pi}\right) = \beta_0 + \beta_1 x_1$$

where $x_1 = 0, 1$. Then

$$OR = \frac{e^{\beta_0+\beta_1\times 1}}{e^{\beta_0+\beta_1\times 0}} = e^{\beta_1}$$

and

$$\beta_1 = \ln OR,$$

providing a very simple relation among the coefficient and the odds ratio. Therefore, the odds ratio is an association measure that indicates how much more likely (or unlikely) it is for the success ($Y = 1$) to be present among those with $X_1 = 1$ than among those with $X_1 = 0$ (Hosmer and Lemeshow, 2000).

For example, if $X_1 = 1$ when the body sensor user heart rate is above some level, $OR = 5$ indicates that a problem is five times more likely to happen when the elderly has that high heart rate.

When we are considering more complex models, the interpretation still is similar for a binary X_i variable. Considering

$$\ln\left(\frac{\pi}{1-\pi}\right) = \beta_0 + \beta_1 x_1 + \ldots + \beta_p x_p,$$

then

$$OR = \frac{e^{\beta_0+\beta_1 x_1+\ldots\beta_1\times 1+\ldots+\beta_p x_p}}{e^{\beta_0+\beta_1 x_1+\ldots\beta_1\times 0+\ldots+\beta_p x_p}} = e^{\beta_i}.$$

For a continuous and linear X_i variable, an increase of c unites implies that

$$OR = \frac{e^{\beta_0+\beta_1 x_1+\ldots\beta_1(x_i+c)+\ldots+\beta_p x_p}}{e^{\beta_0+\beta_1 x_1+\ldots\beta_1 x_i+\ldots+\beta_p x_p}} = e^{\beta_i c}.$$

For instance, if X_i is the user heart rate, an increase of c units on these heart rate originate an increase on the OR of $e^{\beta_i c}$.

For categorical X_i variables, the procedure is similar to the one performed for binary variables. For further details, see Felgueiras (2004).

5 Model Selection

The Elder Care architecture presupposes the production of a large amount of data, since each user should have a body sensor and a location device. One of the main goals of this statistical approach is to select the more important variables that are recorded by the location device and by the body sensor, eliminating redundant variables. Besides, the estimated β_i parameters allow us to assess the importance of the selected variables in foresighting elders problems, because odds and odds ratios can be calculated using the obtained model.

To achieve this goal, a preliminary statistical trial should be implemented within a test group. This test group should be selected using sample techniques, and the body sensors and the location devices should withhold as many information as possible. After the statistical trial, only the selected variables by some model selection method will have to be recorded, allowing a spare in the body sensor and in the location device price.

In this paper we do not refer those sample techniques nor those model selection methods methods, because they are already explained in the several references displayed below.

The variable selection (and the parameter estimation) on a logistic regression model is similar to the variable selection on a linear model. Usual stepwise methods (forward and backward) can be performed, as well as other more complicated methods (Collett, 1999; Hosmer and Lemeshow, 2000).

After a model is obtained, usual validation techniques should be applied before parameter estimates and odds ratio can be interpreted, to assure that the conclusions will be correct.

6 Conclusions and Future Work

The Elder Care solution will allow a 24 hours monitoring for each elderly, leading to a quick response whenever a problem is detected. This solution might be able to increase elderly people quality of life, because survey elders had no longer to abandon their homes. Since real scenario prototype implementation is already beginning, the logistic model will contribute in this first stage to select the more important analyzed variables, and to determine risk factors. After this trial phase, we should be able to have a more simple solution (and therefore cheaper) because unnecessary parameters will no longer be registered. After the comercial implementation of Elder Care solution, the logistic model will be used to classify the elders as having a problem or not having a problem, and therefore this model will be on a constant update to prevent errors connected with sensibility (not detecting a real problem) and specificity (detecting a false problem). This is an important issue, as reducing false alarms reduces sending inappropriate mails and improves the energy efficiency in the network.

However, this effective solution will require a large amount of resources. Therefore, we believe that the architecture presented by Bajo *et al* (2010) can be used in the future with the proposed model, in order to improve the Elder Care solution in terms of resources consumption. The multi-agent system monitors the daily routine, and by using the logistic regression model, presented in this paper, we expect to achieve an high detection ratio (preventing the two types of errors stated before) saving precious computational resources.

References

1. Augusto, J., McCullagh, P.: Ambient intelligence: Concepts and Applications. Comput. Sci. Inf. Syst. 4(1), 1–28 (2007)
2. Bajo, J., Fraile, J., Pérez-Lancho, B., Corchado, J.: The THOMAS architecture in Home Care scenarios: A case study. Expert Syst. Appl. 37, 3986–3999 (2010)
3. Collett, D.: Modelling Binary Data. Chapman & Hall, London (1999)
4. Felgueiras, M.: Análise de Dados Binários. Master Thesys. FC-UL (2003)
5. Hosmer, D., Lemeshow, S.: Applied Logistic Regression. Wiley, New York (2000)
6. Marcelino, I., Pereira, A.: Elder Care Modular Solution. In: Proceedings of The Second International Conference on Advances in Human-oriented and Personalized Mechanisms, Technologies and Services (2009)
7. Marcelino, I., Barroso, J., Bulas Cruz, J., Pereira, A.: Elder Care Architecture. In: Proceedings of the Third International Conference on Systems and Networks Communications (2008)
8. Marcelino, I., Barroso, J., Bulas Cruz, J., Pereira, A.: Elder Care Architecture – a physical and social approach. Adv. Lif. Sci. 2, 53–62 (2010)
9. McCullagh, P., Nelder, J.: Generalized Linear Models, 2nd edn. Chapman & Hall, Boca Raton (1989)
10. Seco, A.: Modelação de Dados Agrupados nas Ciências do Ambiente. Ph.D. Thesys. ISA - UTL (2003)

A New Adaptive Algorithm for Detecting Falls through Mobile Devices

Miguel Sánchez, Patricia Martín, Laura Álvarez, Vidal Alonso,
Carolina Zato, Alberto Pedrero, and Javier Bajo

Abstract. Most of elderly people suffer physical degeneration that makes them particularly vulnerable to falls. Falls cause injuries, time of hospitalization, rehabilitation which is particularly difficult for the elderly and disabled. This paper presents a new system with advanced capacities for learning and adaptation specifically designed to detect falls through mobile devices. The systems proposes a new adaptive algorithm able to learn, classify and identify falls from data obtained by mobile devices and user profile. The system is based on machine learning and data classification using decision trees. The main contribution of the proposed system is the use of posturographic data and medical patterns as a knowledge base, which notably improves the classification process.

Keywords: Context-Aware, Adaptive Algorithms, Machine Learning, Fall detection, Decision trees.

1 Introduction

Falls are one of the most serious problems in the geriatric care because they are one of the fundamental causes of injury and even death in elderly people. According to the World Health Organization (WHO), between 28% and 34% of people 65 and older experience at least one fall per year [14]. In Spain the data are also alarming. Various authors consider that falls are due to a combination of multiple factors, and can be grouped into two main groups: intrinsic and extrinsic factors, and consequences of they are not only physical but also psychological, social and economic [4, 6, 7].

Miguel Sánchez · Patricia Martín · Laura Álvarez · Vidal Alonso · Carolina Zato ·
Alberto Pedrero · Javier Bajo
Universidad Pontificia de Salamanca, c/ Compañía 5, 37002 Salamanca, Spain
e-mail: patrimrodilla@usal.es, miguel@chocosoft.net,
lalvarezba@gmail.com, carol_zato@usal.es,
{valonsose,apedreroes,jbajope}@upsa.es

J.M. Corchado et al. (Eds.): Trends in PAAMS, AISC 90, pp. 17–24.
springerlink.com

Medical studies to identify a patient as an individual propense to suffer falls are based primarily on subjective methods which are difficult to automate or computerize: PCD (Computerized Dynamic posturography) [16, 15], Tinetti test [12, 11] that is based on observable criteria, such as rolling gait, patient's response to an imbalance etc. These methods preset a clear challenge for the computational technologies, not only due to the subjective characters of the indicators, but also because these indicators do not have a standardized treatments. This situation makes necessary to investigate in new solutions aimed at fall's detection in an effective an non-intrusive manner. This paper proposes an innovative adaptive algorithm specifically designed to detect falls. The algorithm obtains data from a detection mechanism built into a mobile phone. The paper proposes an improvement on the traditional methods, incorporating machine learning techniques.

The paper is structured as follows: First we review the state of the art of the technologies involved in the work. Section 3 presents the proposed mechanism to detect falls. Finally, Section 4 presents the results and conclusions obtained.

2 Related Work

Most of the algorithms used for fall detection are based on matrix of data collected by hardware devices, either through specialized sensors or by analysis of images captured by cameras in real time. The data processing is performed in accordance with the type of data being analyzed, but it lacks an intermediate layer able to add metadata to these data. For example, the approach proposed by Y. Zigel [17] is aimed at detecting ground vibration and sound produced by a fall. Another clear alternative is the use of video resources. Rougier and Meunier propose the use of 3D extrapolation of the human head in a video image to recognize a fall [13]. Nowadays, motion recognition using video cameras [8] is widespread, and privacy issues had been resolved by treating images as shapes represented by sets of data [1]. These systems require a complex technological investment to control the environment, and are sensitive to delays in data analysis and false positives. To avoid such problems there are alternative research lines based on data collection through accelerometers and gyroscopes ([2], [3]) that can be integrated in a single device. The data collected corresponds to the parameters of the human movement, different in each individual. In general, with these methods is possible to detect the fall, but not to model correctly the activity and movement associated with it. Chien et al. [5] tried to solve this problem using a k-nearest neighbor classifier that provided good results, but requires complex processing capacities, which are currently very difficult to introduce a standard Smartphone.

Systems based on the study of the relevant data and the noise (which varies depending on the subject and its movement) carry us to the fields of medicine and posturography. The tendency to fall out of elderly people based on their movement had been quantified using computable values in the so-called of Tinetti test, and other papers arising afterwards [9, 10]. Thanks to the work of Raiche et al. [11] and Roqueta [12], we propose to add an intermediate layer based on ambient intelligence in the algorithm. This intermediate layer takes into account the parameters of the patient movement and risk of fall, quantifies a series of patterns and filters the data, separating the valid data from noise. Finally, work on motion analysis according to changes in human posture ([15], [16]) allow us to add more complexity to the proposed algorithm, and to model the human activity of patients with acceptable accuracy. In this sense, the use of input data based on real human motion combined with data obtained from accelerometers and other hardware devices, provides the presented algorithm with advances robustness and efficiency.

3 Proposed Mechanism

This study proposes a new classification method that minimizes the false negatives (those cases in which a fall has occurred and the system does not detects it), and improve the response time. Moreover, the system is integrated within a mobile device, which notably reduces the consequences of falls in patients, reducing the response time to alerts and providing advances information about the patient location. The user wears the mobile device in a waist belt. The mobile device execute a background real time algorithm, which analyzes the data about the user's position and identifies the following situations: standing, sitting, moving or falling, either forward or backward. The proposed algorithm manages two data sources: user profile and data obtained from the mobile device. The following sub-sections explain the data sources and the adaptive algorithm proposed in this study.

3.1 User Profile

The proposed system requires a initial training phase to detect those patient's parameters that influence the Tinetti test [11, 12] and the PCD analysis [15, 16]. A user profile is defined for each patient which contains patterns of his movements as well as information about his disabilities and the risk of falls.

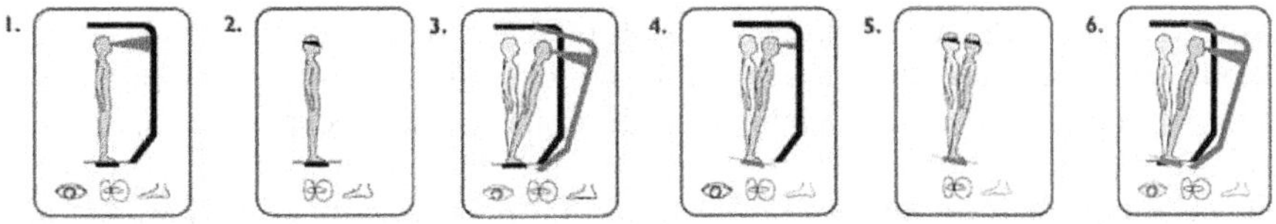

Fig. 1 Sensorial tests proposed in the PCD

As shown in Figure 1, the PCD is based on the assessment of the postural control by studying the movements of the center of pressure (projection of the body center of gravity) under different circumstances in a dynamometric platform. This kind of platform is used to diagnose disorders of the body's balance. This allows the evaluation of the mechanisms that control body stability. That is, it explores the SV sensorial component, vision and proprioceptive receptors, and motor component. The PCD help us to define a series of indicators that are taken into account in the proposed algorithm to quantify the balance and other metrics of the user:

- Degree of stability (Equilibrium score). This parameter defines the stability as a non-dimensional percentage, comparing the maximum amplitude of anterior-posterior swing to the theoretical limits of stability in the anterior-posterior direction. Values close to 100% indicate a minimum swing, while values close to 0% indicates a movement of the center of gravity close to the theoretical maximum of 12.5°. The sensorial analysis can help to detect normal and abnormal patterns. There are four ratios that allow us to detect differences between Condition 1 and the mean score on altered individual conditions.
- Visual ratio (VIS) refers to the ability of the patient to use the visual information. Low values are characteristics of patients who become unstable with uneven support.
- Vestibular ratio (VEST) indicates the ability to use vestibular information. Values close to 0 are obtained for patients which are unstable under conditions with uneven support and without visual system.
- Visual preference ratio (PREF) provides information about the degree of swing with visual reference support. Anomalous values indicative that the patient becomes unstable in visual moving environments.
- Gravity center alignment. Quantifies the location of the patient's gravity center related to the patterns obtained during the training phase. Values out of the central zone determine the limits of the stability.

The PCD and the Tinetti test [11, 12] allow us to provide the proposed algorithm with solid information, to classify falls. Our user profile takes into account three possible groups of users: High, Moderate and Low risk to suffer falls or swing disorders.

3.2 Data Obtained from the Mobile Device

The mobile device must be equipped with a three-axis accelerometer (X, Y, Z) on a single silicon chip. The chip includes all the electronic capacities to process the signals. The accelerometers are now integrated in most mid-range mobile devices on the market. The data taken into consideration are the (x, y, z) coordinates of the patient's movement, the pitch of the terminal, the roll (rotation of the terminal) and the shake typical for a fall. The final algorithm processes these six parameter, together with the user profile and provides classifications, as shown in the following sub-section.

3.3 Proposed Adaptive Algorithm

The proposed algorithm, based on J48 decision trees, is presented in Figure 2.

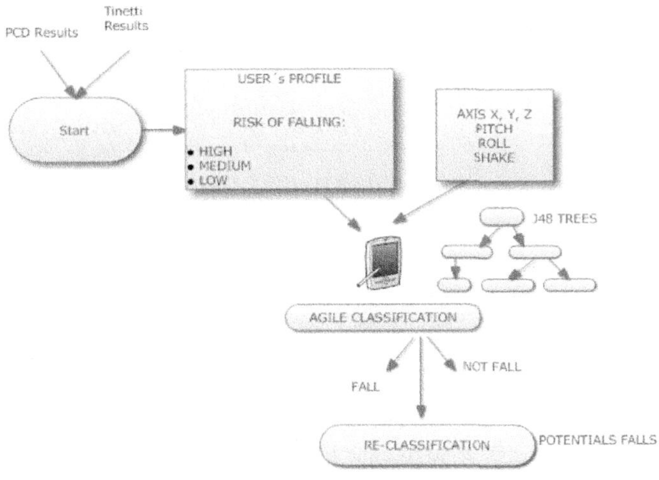

Fig. 2 Algorithm based on J48 decision trees to detect falls via mobile device

As can be seen in Figure 2, in a pre-processing step, non-relevant parameter are filtered. Then, an algorithm based on the J48 decision tree is applied and the situations are classified as FALL, NOT FALL and POSSIBLE FALL. The system provides a real-time response, and the possible falls can be classified using more complex techniques that require extra time. The system is adaptive, since as new data about the patient profile and possible postures, it can learns and adjust the classification parameters that characterize each of the possible status for the patient. In this way, the algorithm allows personalization to the patient profile, adjusting the classification parameters to his particular movements. The choice of classification and learning method was primarily due to the response times and low computational complexity of the decision trees, which are critical when working with resource constrained devices, and to the possibility of removing attributes which are not particularly relevant for the classification or incorporate noisy data. Moreover, decision trees implements a strategy based on the use of the gain ratio criteria, defined as I (Xi, C) / H (Xi). Thus it is possible to avoid potential benefits to the selection of those variables with the greatest number of possible values.

4 Results and Conclusions

The system was tested under simulation conditions. The test's parameters were setted up using previous postural studies related to elderly people. Over three months of testing 500 files were obtained using data from real falls. Initially, the parameters were personalized for the users. Then, the system was tested simulating different fall situations. As shown in Figure 3, the results of classification with three possible outputs in the tree were as follows: 72% of the situations were classified by the decision tree as Fall or Not Tall, and 28% of the situations were

```
Scheme:       .classifiers.trees.J48 -C 0.25 -M 2
Relation:     fallen
Test mode:    10-fold cross-validation
Classifier model (full training set)  J48 pruned tree
zaxis <= 7.504812: FALSE (6.0)
zaxis > 7.504812
|   yaxis <= 0.988897: TRUE (4.0)
|   yaxis > 0.988897
|   |   xaxis <= -1.144109: TRUE (2.0)
|   |   xaxis > -1.144109: FALSE (2.0)
Correctly Classified Instances          71.4286 %
Incorrectly Classified Instances        28.5714 %
Kappa statistic                  0.4167
Mean absolute error               0.2755
```

Fig. 3 Percentages of classification Fall/Not Fall obtained by the proposed approach. The unclassified instances are assigned to the Possible Falls group.

classified Possible Fall (incorrectly classified instances that will be evaluated using a more complex strategy).

The system has a very good response time for those cases that can be identified as Fall and Not Fall. Moreover, as can be seen in Figure 4, it is possible to calculate the cost curve for the misclassification error, providing the system with the ability to detect situations were a fall was detected by error. the system prioritizes the detection of false positives, trying to avoid undetected fall situations.

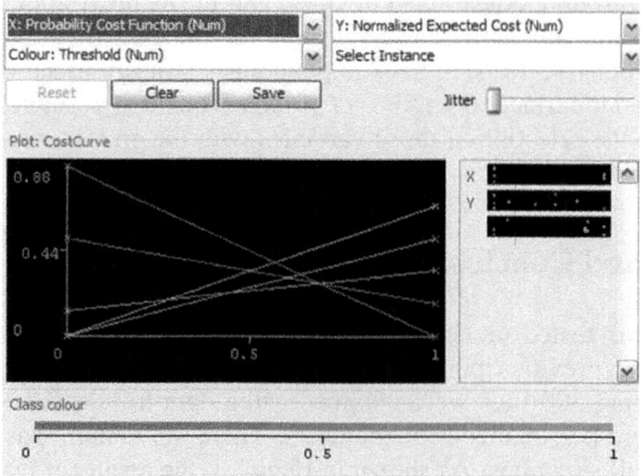

Fig. 4 Cost curve for the misclassification error

The system presented in this study proposes a novel algorithm, that incorporates medical criteria for detecting falls. Current strategies do not incorporate this kind of criteria. The system provides a good response time, since it is able to successfully classify 72% of the cases as Fall or Not Fall using a decision tree-based strategy. The rest of the cases can be identified as suspicious and evaluated using a more computational complex strategy. In adition, the system can be easy integrated in modern distributed architectures, as multi-agent systems, and provides functionalities as alarm systems (using SMS and phone call facilities of mobile devices). The system is able to recalculate the parameters of the user in real-time. The error rate decreases as the system learns from the new input data, and the minimum mean quadratic error obtained in the experiments was 0,16. Our future work focuses on improving the classification algorithm in different ways:

- To improve the accuracy in terms of computerization of PCD and Tinetti indicators in the overall algorithm, reaching even more autonomous and personalized patterns.
- Improve the mechanism to detect the initial position, to facilitate the situation of the device in different parts of the body.
- Improve the sensibility of the parameters of the mobile devices. Effects as shake, can affect the sensibility of the system, and it is possible to automatically tune the parameter and refine the decision boundaries in the decision tree.

Acknowledgements. This study has been supported by the Spanish Ministry of Science and Innovation project TRA2009_0096.

References

1. Anderson, D., Luke, R.H., et al.: others Modeling Human Activity From Voxel Person Using Fuzzy Logic. IEEE Transactions on Fuzzy Systems 17(1) (2009)
2. Benocci, M., Tacconi, C., et al.: Accelerometer-based fall detection using optimized Zig Bee data streaming. Microelectronics Journal 41(11), 703–710 (2010)
3. Bourke, A.K., van de Ven, P., et al.: Evaluation of waist-mounted triaxial accelerometer based fall-detection algorithms during scripted and continuous unscripted activities. Journal of Biomechanics 43(15), 3051–3057 (2010)
4. Carro García, T., Alfaro Hacha, A.: Caídas enel anciano. Residentes de Geriatria Hospital Virgen de Valle. Toledo
5. Liu, C.-L., Lee, C.-H., Lin, P.-M.: A fall detection system using k-nearest neighbor classifier. Expert Systems with Applications 37(10), 7174–7181 (2010)
6. Gama, Z., G6mez, A., Sobral, M.: Epidemiología de caídas de ancianos en España: Una revisi6n sistemática. Rev. Esp. Salud Pública 82(1), 43–55 (2007) (online)
7. Lázaro-del Nogal, M., Latorre-González, G., González-Ramírez, A., Ribera-Casado, J.M.: Características de las caídas de causa neurológica en ancianos
8. Londei, S.T., Rousseau, J., et al.: An intelligent videomonitoring system for fall detection at home: perceptions of elderly people. Journal of Telemedicine and Telecare 15(8), 383–390 (2009)

9. Lustrek, M., Kaluza, B.: Fall detection and activity recognition with machine learning. Slovenian Society Informatika (2009); report of May 2009
10. Nyan, M.N., Tay, F.E.H., Murugasu, E.: A wearable system for preimpact fall detection. Journal of Biomechanics 41(16), 3475–3481 (2008)
11. Raîche, M., Hébert, R., et al.: Screening older adults at risk of falling with the Tinetti balance scale. The Lancet 356(9254), 1001–1002 (2000)
12. de Jaime, C.R.E., et al.: Experiencia en la evaluación del riesgo de caídas. Comparación entre el test de Tinetti y el Timed Up & Go (Experience in evaluation of falling risk. Tinetti and Timed-Up tests comparision). Revista Española de Geriatría y Gerontología 42(6), 319–327 (2007)
13. Rougier, C., Meunier, J.: Demo: Fall Detection Using 3D Head Trajectory Extracted From a Single Camera Video Sequence. Journal of Telemedicine and Telecare 11(4) (2005)
14. Suelves, J.M., Martínez, V., Medina, A.: Lesiones por caídas y factores asociados en personas mayores de Cataluña, España. Rev. Panam Salud. Publica. 27(1), 37–42 (2010)
15. Visser, J.E., Carpenter, M.G., et al.: The clinical utility of posturography. Clinical Neurophysiology 119(11), 2424–2436 (2008)
16. Whitney, S.L., Marchetti, G.F., Schade, A.I.: The Relationship Between Falls History and Computerized Dynamic Posturography in Persons With Balance and Vestibular Disorders. Archives of Physical Medicine and Rehabilitation 87(3), 402–407 (2006)
17. Zigel, Y., Litvak, D., Gannot, I.: A Method for Automatic Fall Detection of Elderly People Using Floor Vibrations and Sound—Proof of Concept on Human Mimicking Doll Falls. IEEE Transactions on Biomedical Engineering 56(12), 2858–2867 (2009)

Agent-Supported Assessment for Adaptive and Personalized Ambient Assisted Living

Helena Lindgren, Dipak Surie, and Ingeborg Nilsson

Abstract. Existing approaches to ambient assisted living (AAL) often fail to consider a human agent's needs from a holistic perspective. In particular the regular assessment of their changing abilities, skills and limitations are often treated as a separate matter in healthcare, thereby affecting the possibilities to provide support tailored to their current condition. Therefore, the objective of this work is to integrate assessment done by the healthcare professional into the framework of AAL. We propose a framework based on a model for activity-centered modeling of knowledge and interaction tailored to users (ACKTUS) and an egocentric interaction model. We exemplify the interaction between system modules as agent dialogues in assessment and adaptation for supporting an older adult in a home environment.

Keywords: personalization; ambient intelligence; smart environments; ubiquitous computing; context-aware computing; pervasive healthcare; agents; assessment.

1 Introduction

In order to provide an individual with ICT-based support in daily living for increasing autonomy, security, health, social inclusion and quality of life, a holistic view on the individuals situation needs to be adopted. The wishes, needs and abilities of the individual need to be assessed in order to be able to optimize the design of the tailored support. Furthermore, the assessments need to be done continuously, in order to adjust the tailored support to changing needs, abilities, wishes and

Helena Lindgren · Dipak Surie
Department of Computing Science, Umeå University, SE-90187 Umeå Sweden
e-mail: {helena,dipak}@cs.umu.se

Ingeborg Nilsson
Department of Social Medicine and Rehabilitation, Umeå University,
SE-90187 Umeå Sweden
e-mail: ingeborg.nilsson@umu.se

J.M. Corchado et al. (Eds.): Trends in PAAMS, AISC 90, pp. 25–32.
springerlink.com

contextual factors. Therefore, we integrate assessment into the framework for developing and maintaining ambient assisted living, with a combination of methods where the professional assessments done by health care professionals plays a key role.

This paper is organized as follows. Section 2 describes the proposed framework and its components that facilitates a holistic approach to AAL, Section 3 presents a scenario in which the proposed framework is explored, developed and evaluated and Section 4 provides conclusions.

2 A Holistic Perspective on Ambient Assisted Living

The framework proposed in this paper synthesizes and builds upon results from 1) a case study where eight older adults were using tailored web interfaces for activity support during two months, where the content and interaction design were based on occupational therapists assessments [1], 2) the work on developing a Semantic Web-based knowledge and interaction modeling environment (ACKTUS) [2], and 3) the work on developing an adopting an ego-centric interaction model for assessing the interaction environment [3, 4]. Prototype systems have been developed and evaluated through a participatory design process involving older adults and experts in different knowledge domains. These prototype systems will in our framework collaborate by means of a service-oriented architecture through the proposed multi-agent system and constitute an ambient assisted living (AAL) home environment that serves the purpose to support and maintain a holistic user-centered view of an individual's life situation. The tailoring of the AAL environment to an individual is ideally based on a combination of automated methods for skills and ability detection, activity recognition and evaluation with the professional's regular assessment of the individual's ability, needs and wishes to perform different activities (e.g., ADL and leisure assessments [5, 6]). The professional's assessment provides information about how interactive systems should be adapted to the individual, e.g., based on physical, cognitive, social and/or psychological limitations and abilities. The common domain knowledge and assessment instruments are currently being modeled by domain experts using ACKTUS (a Semantic Web application for Activity-Centered modeling of Knowledge and interaction Tailored to Users [2]). Our approach supplements other approaches (a summary of interesting approaches is given in [7]) in that it is both preventive and integrates AAL into a context of assessment, intervention and follow-up in daily practice at the point where new knowledge is generated and integrated. This makes our approach easily adapted to individuals and extended with new evidence-based or best practice knowledge, models for interpretation, evaluation, etc. as a natural part of daily practice and activity.

In this section we describe the following components of our framework: the users, personalized applications with tailored interaction design and content, smart everyday objects, and give an overview of the agent design, agent interaction and communication (Figure 1). An initial technical description of the argument-based inquiry dialogues used in our example scenario is provided in [8].

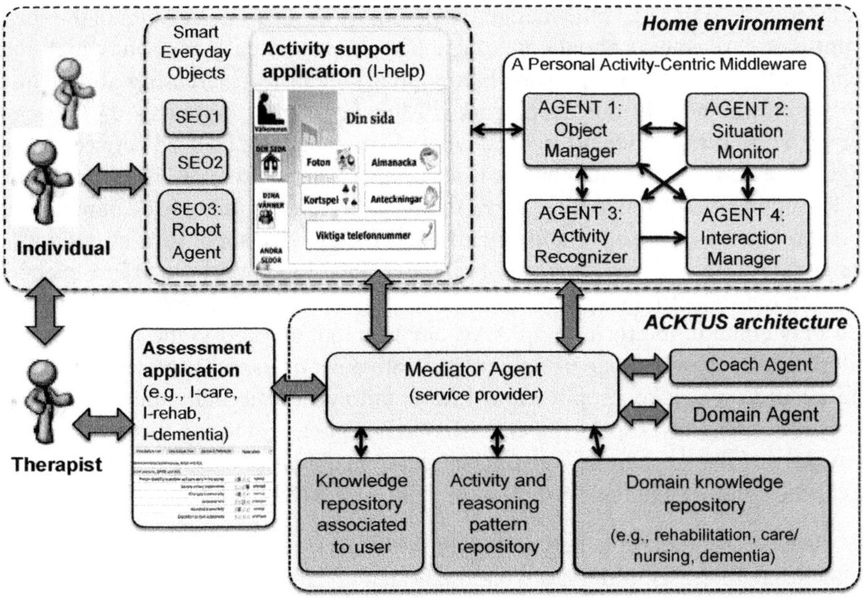

Fig. 1 Overview of the use environment of the proposed framework

We model two types of users (human agents), the individual in focus and other actors in the home environment, and the therapist (e.g., occupational therapist (OT) or other professional). For each type of user there is a dedicated and personalized ACKTUS application: 1) I-help for supporting the individual and relatives in activity (e.g., [1]), and 2) I-rehab for supporting the OT in daily work, I-care for the nurse and I-dementia for the physician. The different applications provide personalized interactive views of the relevant domain knowledge and actor knowledge, tailored to the user based on the user profile, modeled in the actor ontology and stored in the user knowledge repository (Figure 1) [2].

The applications are developed using ACKTUS, which is a semantic web environment that is being developed for modeling knowledge to be integrated in support systems in health care, and for designing the interaction with the end user applications [2]. ACKTUS consists of a service-oriented architecture, which includes an RDF/OWL ontology that captures 1) components of reasoning in the form of an argumentation framework, 2) components used for tailoring interaction with the resulting knowledge applications and 3) components for modeling the user agents as actors in a situation where the application is used, e.g., for reasoning, performing daily activities, for entertainment or social interaction. The ACKTUS applications share this common base ontology, which is extended with specifics for each knowledge domain [2]. Thus, for each knowledge domain (care, rehabilitation, etc.) a domain knowledge repository is created in the modeling of knowledge, containing validated assessment instruments (e.g., [5, 6]), clinical practice guidelines, etc. (Figure 1). Preference orders among knowledge sources, level of expertise in knowledge domains and other contextual factors are taken into account in the (partly agent-based) reasoning about e.g., diagnosis in a patient case or about which interventions to be made in a persons home environment for increasing ability in activity performance. In our framework a Mediator Agent organizes the dialogues between different application agents, where knowledge can be used between knowledge repositories, depending on the topic to investigate and on users' authorization.

The application I-rehab is focussed in this paper since it is being designed to support the OTs assessment of ability and need for activity support in an individuals life situation. In addition, it is also being designed to support the OT in executing interventions according to goals formulated together with the individual. Intervention may come in the form of an AAL environment tailored to the individual. The tailoring may then be done using I-rehab, with a set of useful applications to adopt and adapt in a mockup application, I-help, tailored to the individual (a prototype is visible in Figure 1). Apart from a Mediator Agent, which functions as a service provider and does not contribute to or affect the content of dialogues, the following agents are activated in the interaction with an ACKTUS application (Figure 1):

1. The Coach Agent (CA) acts on behalf of the user, represents the user and guards the users interests. It detects what the human user agent does in interaction with the application, organizes and updates the users belief base, and acts when the user does not follow previously stated preference orders or pattern of interaction (partly represented in the users belief base). The CA also organizes the user model, which collects the information about the users limitations and resources (e.g., physical, cognitive, social) and preferences about e.g., who has access to the information. The CA detects pattern of activity as well as deviations from pattern of activity in an individual user based on information from the personal activity-centric middleware [9].
2. The Domain Agent (DA) acts as a domain professional, making use of the domain knowledge repository. In the interaction with the CA and the therapist user in the assessment application, the DA provides the expert perspective to reasoning, when possible based on evidence-based guidelines and validated assessment instruments for the purpose to educate the therapist and support daily work with e.g., older persons who need adjustments in their home environment. DA represents the therapist in the interaction with the CA, which in turn represents the individual, or with the individual him or herself through the I-help application.

As a supplement to the explicit knowledge and interaction modeling, automated methods for skill and ability detection, activity recognition and activity evaluation can be integrated (e.g., [1, 3]). While the individual can through interaction with a tailored support system, give more or less explicit views on his or her situation, needs, motives, wishes, etc., the system could also automatically detect deviations from a state with certain interaction pattern, which may be caused by a development of cognitive deficits or other health conditions. The home environment of an individual is expected to be populated with everyday objects augmented with mediators like sensors, actuators, interactive devices, user interfaces, artificial agents, etc. which we refer to as smart everyday objects (SEOs). The individual accesses the I-help application both explicitly and implicitly through the SEOs. The personal activity-centric middleware for detecting activities and distributing interaction over SEOs in a home environment [9] includes the following (Figure 1):

1. The Object Manager (OM) (Agent 1) is responsible for wireless communication with SEOs populated in the individual's environment through WLAN, Zigbee and Bluetooth communication protocols depending on the SEO capabilities and the data exchange requirements. The OM keeps track of the identity of the SEOs that are available for interacton with the individual in their environment, and the actual interaction performed by that individual with SEOs resulting in state changes to the SEOs. This includes both physical (e.g. Turn-off a stove) and virtual (e.g. Send an email) state changes to the SEOs, and both internal (e.g. Turn-off a stove) and external (e.g. Place sauce pan on the stove) state changes to the SEOs. The object manager maintains a database about the available SEOs, their capabilities, and upgrades such information over time by communicating with the SEO manufacturers similar to online software updates.
2. The Situation Monitor (SM) (Agent 2) is responsible for monitoring the individual's situation using a Situative Space Model [4], conceptually based on the individual's ability to perceive and act in an environment. The set of objects that are perceivable and the set of objects that are actable by a human agent at a particular instance in time is framed using the wireless signal strength measures (RSSI) at the SEOs in reference to the individual and the SEO properties. The term object refers to both physical (e.g. a tap) and virtual (e.g. a chat window) objects thereby keeping track of the individual's physical and virtual situation.
3. The Activity Recognizer (AR) (Agent 3) is responsible for modeling and recognizing the individual's activities and actions [3]. The individual's activities are modeled using parallel Hidden-Markov Models (HMMs), where each activity is modeled using a parallel HMM with its states representing the actions within the respective activity. A human agent's interaction and manipulation of SEOs is fed as input to the parallel HMMs by communicating with the OM agent, and the human agent's situation is also fed as input to the parallel HMMs by communicating with the SM agent. Activities like cooking lunch, washing clothes, paying bills through the internet, etc. are detected by the AR, which are recognizable by the CA as activities specified by the individual and formally organized in the activity ontology and the CA's belief base.
4. The Interaction Manager (IM) (Agent 4) is responsible for facilitating the individual's interaction with I-help through the SEOs present in the individual's environment [9]. If I-help requires explicit input from an individual, it presents relevant information through the SEOs that are potentially perceivable by that individual inducing him or her to act on it or any other SEO that is currently actable in the environment. The IM agent communicates and coordinates with other middleware agents in deciding on what, when and how to select an appropriate SEO for exchanging virtual information in the AAL environment. The IM agent uses simple rules that consider the individual's current activity and action context, their current perceptual and action possibilities, available situative SEOs and the information exchange requirements imposed by I-help.

The activity-centric middleware is implemented in C$^\sharp$, while the communication between the middleware agents is handled using Microsoft Messaging Queues (MSMQ) that sends and receives information reliably using a temporary storage across ad-hocly networked computers. The CA and DA agents follow a protocol for argument-based inquiry dialogue for collaborative knowledge building and problem solving. The formal protocols for dialogues are adopted from [10] and are extended with values and preferences drawn from the ACKTUS argumentation framework and knowledge repositories [8]. The purpose of the agents' dialogues is to share (defeasible) knowledge to construct arguments supporting beliefs, and in this way increase each agents knowledge repository (belief base). We assume that the participants are cooperative and reliable, aiming at disseminating and increasing knowledge and finding optimal decisions and actions. We assume also that one single medical actor does not possess all knowledge required for providing a client optimal care, but that the knowledge is distributed over a team of professionals (and corresponding knowledge repositories) with different viewpoints of a situation.

The dialogues can be activated by any of the agents, and the content of messages is selected from their belief bases, based on the topic for the dialogue [8]. The topic is framed by the underlying ontology, which defines the reasoning context and associated critical questions, rules, knowledge sources, etc. [2]. Critical questions function as locutions in dialogues. A dialogue can also involve the human agent, then visible through the I-help application and possibly through audio modality. The setup for the dialogue is arranged in collaboration between the CA and the IM, where the CA mediates preferences on how a dialogue initiated by the CA should be arranged with respect to modality, medium and situation. The way the human agent can initiate a dialogue is also specified. A topic needs to be selected, possibly among predefined topics visible through a graphical user interface, or by vocal interaction.

3 The Grandma Rut Scenario

Our example is an authentic case who suffered from a few falls before a hip fracture became the result with a long hospitalization with severe anxiety and fatal complications as a consequence. We envision a different scenario by introducing the AAL environment as a supplementary intervention to better meet her changing needs. This case shares needs and wishes also with some of the participants in an earlier case study and is therefore considered representative [1]. Our example is a woman, Rut who is 84 years old, who suffers from pain in her back and legs, which prevents her from walking out from her apartment on her own. She also needs a walking aid in her home. The OT has visited her home, made interviews and assessments about needs of adjustments in the home using I-rehab, which includes domain-specific assessment instruments. Ruts son was also present to tell his view on Ruts life situation. The assessment shows that Rut has a mild episodic memory deficit and possibly other mildly decreased cognitive functions, and has a mild depressive state with worries about falling and not being able to get help. Therefore, she is becoming more and more passive and isolated. She wishes to have more contact with her friends and children. She does not feel that the memory deficit affects her ability.

Her son has a different view, and has noticed increasing difficulties in remembering taking medication and eating meals.

Based on goals for interventions formulated together with Rut and her son, the OT creates an I-help profile for Rut with the assessment information about activity, participation, cognitive ability, interests, etc. as underlying the application, and designs the content by selecting appropriate functionalities. In addition to integrated SEOs in the environment, a service robot has moved in, which she calls Butler.

Eight months later, the CA notices a change of pattern in Ruts activity. She has begun to visit the bathroom a couple of hours after she has went to bed. The CA has older information that Rut sleeps well the whole night with the help from the sleeping pills she gets. This contradictory information together with the knowledge about Ruts fear of falling when walking makes the CA initiate a dialogue with the DA with the topic *risk for falling*. The DA has knowledge that walking around during night increases the risk for falls and activates in turn another dialogue to assess the risk situation with the DAs for the care repository and dementia repository. The following possible causes for the deviation of sleeping patterns are identified based on knowledge from the different knowledge repositories: 1) the sleeping pills do not work, 2) possible incontinence condition, 3) possible decrease in cognitive ability related to time and space perception, 4) increase in pain condition and/or 5) an increase in worries and anxiety. The CA is obliged to resolve the conflicting belief base and initiates a dialogue with Rut next morning about sleep with Butler as medium. The following question is directed to Rut: *Did the sleeping pill work for you tonight?* If Ruts answer is no, the following question is given: *Do you want to discuss the pills and your sleep with the nurse?* If the answer is yes, the agent tells Rut that the nurse is informed and may contact Rut. In order to complete the assessment about risk for falling, the CA may continue the dialogue with Rut through Butler about circumstances in her daily life based on questions defined in the protocol for risk assessment that are to be answered by the person at risk. When the agent has done the assessment to the extent it is able to, Rut is shown a summary and Butler asks if it is OK to show it to the nurse. If she says yes, the nurse is able to use the information in the preparation for a visit. During a visit the nurse investigates the pain condition, mental health, a possible incontinence, and completes the assessment about risk for falling. If Rut is walking about at nighttime, Butler is alerted to be on guard for detecting sounds that may indicate a fall event, or to react if Rut spends longer time than usual in the bathroom. In these cases, Butler can ask Rut if she is okay. If she does not answer, or says no, Butler can act based on the information (e.g., ask if it should call her son, or send an alarm to care personnel).

To summarize our example, the main benefits of introducing the ambient assisted living framework in Ruts case is the potential decrease of fear of falling and its consequences, increase sense of security, e.g., having the service robot paying attention by asking how she is doing, and increasing the feeling of being part of a social context by providing an easy way to make contact with, and having an open channel of interaction with people outside her home in her daily activities. Another benefit is also the important possibility to detect of deviations from typical activity pattern that may be caused by an evolving dementia disease.

4 Conclusions

A framework for an agent-based AAL is described, which is synthesizing earlier work for the purpose to provide a holistic perspective on the user's life situation. The approach supplements other approaches to AAL in that it integrates assessment done by care professionals as a key feature. The assessment is used for goal setting for interventions, which may partly come in the form of a tailored ambient assistant home environment. Initial and continuous assessments provide also baselines for follow-up on and adjusting the environment, in order for the system to adapt to changing ability and needs in the individual. In this iterative assessment and adaptation process agents with different roles and with access to different knowledge repositories may resolve issues in dialogues between the participating actors. Ongoing work includes the implementation of the agent dialogue protocols and the modeling of the rehabilitation and care domain knowledge.

References

1. Lindgren, H., Nilsson, I.: Designing Systems for Health Promotion and Autonomy in Older Adults. In: Gross, T., Gulliksen, J., Kotzé, P., Oestreicher, L., Palanque, P., Prates, R.O., Winckler, M. (eds.) INTERACT 2009. LNCS, vol. 5727, pp. 700–703. Springer, Heidelberg (2009)
2. Lindgren, H., Winnberg, P.: A Model for Interaction Design of Personalised Knowledge Systems in the Health Domain. To appear in Proc. 5th International Workshop on Personalisation for e-Health (2010)
3. Surie, D., Pederson, T., Lagriffoul, F., Janlert, L.E., Sjölie, D.: Activity Recognition using an Egocentric Perspective of Everyday Objects. In: Indulska, J., Ma, J., Yang, L.T., Ungerer, T., Cao, J. (eds.) UIC 2007. LNCS, vol. 4611, pp. 246–257. Springer, Heidelberg (2007)
4. Surie, D., Jäckel, F., Janlert, L.E., Pederson, T.: Situative Space Tracking within Smart Environments. In: Proc. of the 6th International Conference on Intelligent Environments, pp. 152–157. IEEE Computer Society Press, Los Alamitos (2010)
5. Ejlersen Wæhrens, E.: Measuring quality of occupational performance based on selfreport and observation: development and validation of instruments to evaluate ADL task performance. Umeå University, Doctoral thesis (2010)
6. Nilsson, I., Fisher, A.G.: Evaluating leisure activities in the oldest old. Scand. J. of Occupational Ther. 13, 31–37 (2006)
7. Kamel Boulos, M.N., Lou, R.C., Anastasiou, A., Nugent, C.D., Alexandersson, J., Zimmermann, G., Cortés, U., Casas, R.: Connectivity for healthcare and well-being management: examples from six European projects. Int. J. Environ. Res. Public Health 6, 1947–1971 (2009)
8. Lindgren, H.: Towards Context-Based Inquiry Dialogues for Personalized Interaction. To appear in Proc. 9th International Conference on Practical Applications of Agents and Multi-Agent Systems (2011)
9. Surie, D., Pederson, T., Janlert, L.E.: The easy ADL home: A physical-virtual approach to domestic living. J. of Ambient Intelligence and Smart Environments 2-3, 287–310 (2010)
10. Black, E., Hunter, A.: An inquiry dialogue system. Autonomous Agents and Multi-Agent Systems 19(2), 173–209 (2009)

A Multi-Agent System for Remote Eldercare

Boštjan Kaluža, Erik Dovgan, Violeta Mirchevska,
Božidara Cvetković, Mitja Luštrek, and Matjaž Gams

Abstract. This paper presents a case study in which a multi-agent system for care of the elderly people living at home alone is applied in order to prolong their independence. The system consists of several agents organized in groups providing robust and flexible monitoring, calling for help in the case of an emergency and issuing warnings if unusual behavior is detected. The first results and demonstrations show promising performance.

1 Introduction

Population aging in developed countries presents an increased pressure on the working-age population as well as on the national health budgets. A lot of elders cope with at least one health problem and many are dealing with two or more, e.g., hearth problems, arthritis, osteoporosis, cancer, frequent falls (which can lead to fractures), etc. Faced with decreased ability to function, they require either additional help or move to an eldercare facility. However, given the choice, elders mostly prefer to continue to live at home instead in an eldercare facility.

In this paper we present a case study of a multi-agent system that aims to prolong the independence of the elderly people living alone at their homes. The system is part of the Confidence project [5], which main objective is to build a remote care system for the elderly by detecting (i) short term disabilities such as falls and (ii) long term disabilities such as development of a disease. This paper presents the multi-agent architecture of the system and describes its prototype implementation.

The paper is organized as follows. Section 2 first presents a sample scenario as well as our vision how technology can promote confidence to both elders who live alone and their concerned family members, and second, it reviews the related

---

Boštjan Kaluža · Erik Dovgan · Božidara Cvetković · Mitja Luštrek · Matjaž Gams
Jožef Stefan Institute, Department of Intelligent Systems, Jamova 39,
1000 Ljubljana, Slovenia

Violeta Mirchevska
Result d.o.o., Bravničarjeva 11, 1000 Ljubljana, Slovenia
e-mail: {bostjan.kaluza,erik.dovgan,violeta.mircevska,
boza.cvetkovic,mitja.lustrek,matjaz.gams}@ijs.si

J.M. Corchado et al. (Eds.): Trends in PAAMS, AISC 90, pp. 33–40.
springerlink.com

work. Section 3 is devoted to description of the multi-agent architecture and its components, while Section 4 describes prototype implementation and results. Finally, the last section outlines future work.

2 Remote Eldercare

2.1 Scenario

Mrs. Brown, aged 71, lives in her house by herself since her husband passed away. Her grownup children moved to a different city and are able to visit her once or twice a month. Recently, Mrs. Brown tumbled on the floor while trying to clean the cabinet. She broke her left hip. After the recovery, she is able to prepare simple meals by herself, get dressed and take a bath or shower, but she feels unsecure and has an increased fear of falls. She avoids cleaning and walking around the house, and complains that she often feels dizzy. Although her doctor and children suggested several times, Mrs. Brown does not like the idea of moving from her home to an eldercare facility. In this situation, Mrs. Brown is living alone and despite not being at risk in terms of getting medical assistance there is a need for someone to take a look after her.

This aspect can be improved using remote eldercare technology, namely ambient intelligence including sensors, context-aware reasoning and on-line services. Now, let us imagine an updated scenario in which Mrs. Brown lives alone in a smart house under the care of the Confidence system. Her house is equipped with wireless sensors to monitor and analyze her posture and activities of daily living (ADL). The Confidence system is deployed to analyze her walking characteristics and detect any changes that might indicate a decreased performance, either related to her recent hip fracture or development of a new health problem. Confidence also monitors her activities of daily living in order to highlight deviations possibly correlated with illness or some other disability. In case Mrs. Brown suffers from a fall or just does not feel well, the Confidence system is trained to recognize even complex emergency situations and to call relatives or emergency center if needed. Overall, Mrs. Brown is still in her house, confident that if something goes wrong she will receive help even if she won't be able to make a call. Her concerned family members and doctor are now convinced that she can stay at home.

The primary goal of this research is to create a remote care system that can help enhance the quality of care for elders living in a smart environment by providing (i) detection of abnormal situations, meaning that the system is able to detect abnormal situations in short-term behavior, such as falls and loss of consciousness (in this case the system raises an alarm that results in phone call to family members or emergency center), and (ii) monitoring of behavior changes over longer periods of time in order to detect changes that indicate decreased performance. Gait analysis, for example, may discover gait disorders, which are an important indicator of a number of health problems: joint and muscular diseases, neurological disorders such as Parkinson's disease etc. Moreover, the goal is also to notify an increased danger of falling, which may appear with time, so that the user can start walking with a cane or a walker.

2.2 Related Work

Fall detection is one of the major challenges in eldercare, since falls are usually the main cause of hospitalization: older adults are five times more likely to be hospitalized for a fall-related injury than for any other cause [16] and moreover, fall injuries account for 40% of all nursing-home admissions [7]. Several systems were introduced to tackle fall discovery with accelerometers [6], [11], [19] and gyroscopes [2]. Pan et al. [15], for example, presented a homecare service for fall discovery using a body-worn tri-axial accelerometer and reporting such a discovery to an emergency center. Also, accelerometer-based fall detectors are already commercially available, for example, iLife™ [1] that employs panic button, accelerometers to detect falls and wireless transmitter, and Zenio [18], that provides similar product equipped with Bluetooth, so it can be connected to a PC or a mobile phone.

In recent years, agent-based systems have gained great importance in the healthcare field. Isern et al. [8] reviewed several agent-based solutions concluding that such an approach has positive affects in terms of modularity, efficiency, decentralization, flexibility, personalization, distributed planning, monitoring, proactivity and security. Luis et al. [3] presented the TeleCARE system, a multi-agent platform for virtual communities in elderly care. The main idea is to provide a framework upon which a variety of services can be employed. In addition, the project RoboCare [4] aims to develop a system consisting of a set of robotic and software agents to control the environment. The system tracks the elderly person with a camera, controls his or her daily activities and reports on any inconsistency in the execution.

Compared to the agent-based systems, which basically dealt with just one task, the objective of our research was twofold. The primary goal was to create a care system that is able to detect hazardous situations in short-term behavior (falls, loss of consciousness), which is one of the most common tasks in eldercare. The second, rarely addressed goal was to detect changes in behavior over longer periods of time. An important advantage of the system is that it is based on localization hardware (that enables context-depended reasoning resulting in a lower false-alarm rate). Localization enables a crucial advantage over accelerometer-based input because knowing that a person is in a bed enables avoiding annoying false alarms even when fast accelerations occur.

3 System Architecture

The system is designed as a multi-agent system. Each system module, task or activity is designed as an agent, providing a service. Agents are organized into groups at a specific level of abstraction and coordinated by another, hierarchically higher-level agent. Each agent can be simply modified or replaced and new or redundant agents can be easily incorporated. An agent provides a service that is requested by another agent or is triggered by some event. Agents share their data through direct acquisition using three types of messages:

1. **Measurement message** is created when a new measurement is obtained from sensor agents. In this stage it contains only raw sensor data. The message is later augmented by other agents with additional interpretations, e.g. filtered/derived data, posture information, etc.
2. **Action message** is generated when an agent or an agent group requires service from another agent or group. Typical scenario is when a group of agents detect a fall and require alarm action from the communication agent group.
3. **Status message** is used to pool or post agent state.

The MAS architecture is illustrated in Figure 1. The main groups of agents and their interactions are presented with arrows.

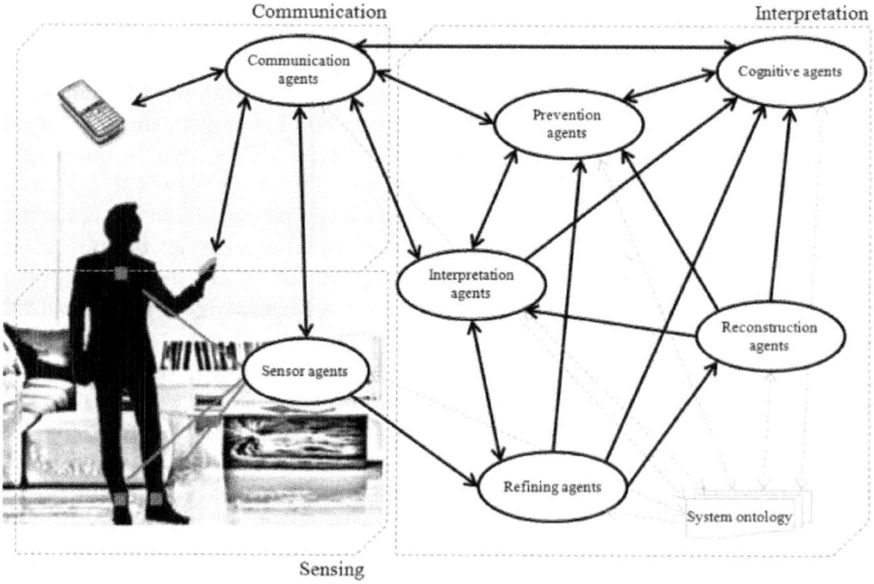

Fig. 1 Multi-agent architecture

The architecture presented in Figure 1 contains seven groups of agents. Each group can contain several agents performing the same task (e.g., reading sensor data, recognizing posture) and some agents that coordinate final result of the group. At the lowest layer, an arbitrary location hardware system is employed. Sensing agents read raw sensor data, usually in sequential time periods (1/10 second), and serve them in a form of measurement message. At the end, a synchronization agent merges measurement messages within given time stamp to a single message. The sensor agents can also report other information, for example, a status message "low battery" is sent to communication agent group, which redistributes message to user and graphical interface.

The refining agents filter the noise, compute the derived attributes and map the raw data with the human-body model. As a result, the refining agent group augments the measurement message by providing a uniform representation of all available data of the body.

The reconstruction agents aim to determine the posture of the person in the environment. The group consists of classification agents based on machine learning (Random Forest) [12] and expert rules [14]. Coordination agent pulls a measurement message from the refining agent group and redistributes it to the classification agents. When all classification agents provide their label for the posture, the meta-classification agent merges (implemented with HMM) labels to the final posture classification.

The interpretation group of agents tries to detect if a person is in a dangerous situation. Consider the following situation as an example: an agent detects that the user is immovable, the reconstruction agents provide posture that the user is lying, and the refining agents give location 'kitchen'. The interpretation agents, implemented with data-driven (SVM) and knowledge-driven approach (expert rules), classify this situation as risky, since it is very unusual and most likely related to a health problem, e.g., the person might have lost consciousness, and inform the communication agent group.

The prevention-agent group monitors how the person behaves at various levels [13]. It consists of several agents observing a variety of statistics, e.g., walking characteristic, activity characteristics, daily dynamics [9], ADL, etc. Each agent pulls relevant measurement messages from other agents and builds its own behavior model (implemented with outlier detection). When an agent detects a deviation, it notifies group coordination agent, which decides upon notifying communication agents.

The next group is cognitive-state agent group: increasingly more abstract context awareness makes it possible to construct the cognitive state of the integrated system. The system design includes the cognitive state of the user, although not implemented yet in the tested version (all the other levels are already implemented). The goal is to reconstruct not only the physical, but also the cognitive state of the user and perform reasoning on a wider spectrum of information with an integrated reasoning strategy.

The last group consists of communication agents that are dedicated to user interaction, for example, agents that alert the user with a reply demand, make a phone call to relatives or help center, graphically display the state of the system, etc. An example of the communication agent is an alarm agent that triggers when a dangerous situation is reported by other agents and shuts down when the user responds that everything is OK.

4 Demonstration

Our demonstration environment (a room) was organized as a near-realistic home apartment, equipped with a bed, a few chairs and tables, and divided into five logical areas: a kitchen; a sleeping area, a living area; a toilet; and a corridor. As localization component we selected the commercially available system Ubisense

[17] that allows local positioning by tracking a set of small tags worn by a user. The tags were placed on chest, belt (optional), left and right ankles.

We have implemented a set of communication agents and graphical interface as depicted in Figure 2. The first three frames display the state of the reconstruction agent group: top perspective of the user in the room (frame 1); side view as coordinates x and z (frame 2); and the recognized activity by particular agents and joint prediction (frame 3). Next, the state of the interpretation-agent group is displayed in frame four with classification of situation. Frame five is dedicated to a live video stream (only for testing purposes, not part of the system). The agents from prevention group are displayed with histogram of movement (frame 6), a visualization of the daily-dynamic-analysis agent (frame 9) and other agents (frame 10). The status of the sensor agents and the available tags is presented in frame 7, while the user-interaction agent (to cancel the raised alarm or trigger an alarm manually) is in frame 12. Currently, our prototype implementation is PC-based, while we are in the process of extending it with more advanced service-based agents such as an emergency call service, a portable device for alarm inhibition, a mobile phone, a remote web-based monitoring service, etc.

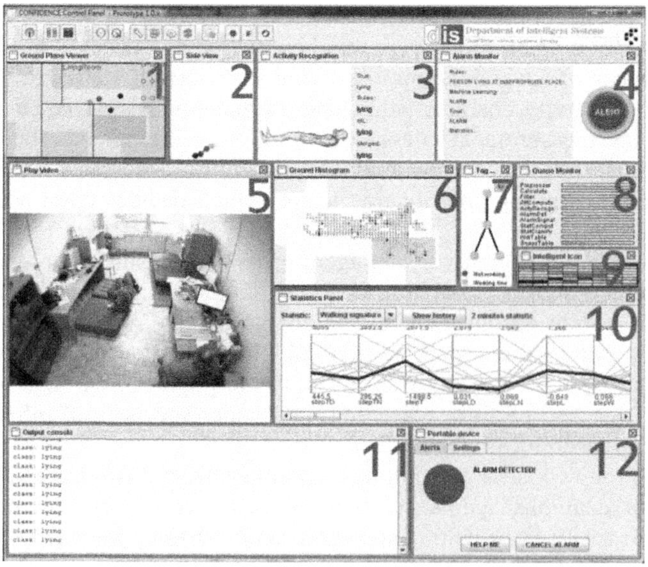

Fig. 2 The communication and graphical interface of the prototype system for eldercare

The system has been intensively tested in several complex scenarios where alarm must be raised and situations where false alarm is likely. The testes were performed in our testing room with young volunteers simulating walking, sitting, falling etc. The tests presented in [10] are shown in comparison with accelerometer-based system only and accelerometers with body orientation. The results in Table 1 show a significant difference, but one should have in mind that the cases were chosen to deal with situations that are hard to classify. More precisely, in a

Table 1 Performance evaluation of the Confidence system and systems using accelerometers

Case / Accuracy [%]	Confidence system	Accelerometers and orientation	Only accelerometers
A: Tripping	100.00	100.00	100.00
A: Fainting standing	100.00	0.00	0.00
A: Sliding from the chair	52.00	0.00	50.00
FA: Jumping in bed	100.00	0.00	0.00
FA: Sitting down quickly	100.00	100.00	0.00
FA: Searching for something under table/bed	96.00	100.00	100.00
Overall	**91.33**	**50.00**	**41.67**

typical test, only simple laboratory cases are presented and publications typically report about 100% accuracy. However, in real life, 100% accuracy is not achieved. This is also the case with the Confidence since it was not able to detect with high accuracy if a person is sitting on a chair or on a floor in a time limit of 10 seconds. This is due to inaccurate position data that have standard deviation around 20 cm. However, tripping, i.e., falling from walking, and fainting from standing, i.e., slowly lying down when fainting, were captured with 100% accuracy. Accelerometers with or without orientation are successful when dealing with fast falls, however, fail to recognize lying down slowly due to fainting and sliding from the chair. The last three rows in Table 2 denote cases when no alarm should be reported. In these cases, accelerometers with orientation perform better than accelerometers without orientation while the Confidence performs close to 100%. Demo of this particular experiment with videos and results is available on http://dis.ijs.si/confidence/demo.html.

5 Conclusion

In this paper we have presented a multi-agent system for remote eldercare that consists of several agents organized in groups that are able to detect falls and other situations potentially dangerous to the person. When compared to the accelerometers, the system achieved over 40% better performance. The Confidence project is in its final year, and is mainly devoted to in-depth testing and verification. However, there are a couple of directions of current interest, e.g., the initialization procedure, user-friendly communication, and the cognitive-agent group.

Acknowledgments. This work was partly supported by the Slovenian Research Agency under the Research Programme *P2-0209 Artificial Intelligence and Intelligent Systems*, and partly from the EU's FP7/2007–2013, grant agreement No. 214986. The work of Violeta Mirchevska was partly financed by the European Union, European Social Found. We would like to thank Domen Marinčič, Rok Piltaver, Blaž Strle, and Damjan Kužnar for suggestions and discussions.

References

[1] AlertOne Services, Inc. iLife™ Fall Detection Sensor, http://www.falldetection.com

[2] Bourke, A.K., Lyons, G.M.: A threshold-based fall-detection algorithm using a bi-axial gyroscope sensor. Medical Engineering & Physics 30(1), 84–90 (2008)

[3] Camarinha-Matos, L.M., Afsarmanesh, H.: A multi-agent based infrastructure to support virtual communities in elderly care. Int. Jour. of Networking and Virtual Organisations 2(3), 246–266 (2004)

[4] Cesta, Pecora, F.: Integrating Intelligent Systems for Elder Care in RoboCare. In: Mann, W.C., Helal, A. (eds.) Promoting Independence for Older Persons with Disabilities, pp. 65–73. IOS Press, Amsterdam (2006)

[5] Confidence (April 23, 2010), http://www.confidence-eu.org

[6] Doughty, K., Lewis, R., McIntosh, A.: The design of a practical and reliable fall detector for community and institutional telecare. Journal of Telemedicine and Telecare 6, 150–154 (2000)

[7] Fuller, G.F.: Falls in the elderly. American Family Physician 61(7), 2159–2168, 2173–2174 (2000)

[8] Isern, D., Sanchez, D., Moreno, A.: Agents applied in health care: A review. International Journal of Medical Informatics 79(3), 145–166 (2010)

[9] Kaluža, B., Gams, M.: An Approach to Analysis of Daily Living Dynamics. In: Proceedings of the WCECS 2010, ICMLDA 2010, San Francisco, CA, vol. 1, pp. 485–490 (October 2010)

[10] Kaluža, B., Mirchevska, V., Dovgan, E., Lustrek, M., Gams, M.: An Agent-based Approach to Care in Independent Living. In: de Ruyter, B., Wichert, R., Keyson, D.V., Markopoulos, P., Streitz, N., Divitini, M., Georgantas, N., Mana Gomez, A. (eds.) AmI 2010. LNCS, vol. 6439, pp. 177–186. Springer, Heidelberg (2010)

[11] Kangas, M., Konttila, A., Winblad, I., Jamsa, T.: Determination of simple thresholds for accelerometry-based parameters for fall detection. In: Proceedings of the 29th Conference of the IEEE Engineering in Medicine and Biology Society, pp. 1367–1370 (2007)

[12] Luštrek, M., Kaluža, B.: Fall detection and activity recognition with machine learning. Informatica 33(2), 197–204 (2009)

[13] Luštrek, M., Kaluža, B., Dovgan, E., Pogorelc, B., Gams, M.: Behavior analysis based on coordinates of body tags. In: Tscheligi, M., de Ruyter, B., Markopoulus, P., Wichert, R., Mirlacher, T., Meschterjakov, A., Reitberger, W. (eds.) AmI 2009. LNCS, vol. 5859, pp. 14–23. Springer, Heidelberg (2009)

[14] Mirchevska, V., Luštrek, M., Gams, M.: Combining machine learning and expert knowledge for classifying human posture. In: Proceedings of the Eighteenth International Electrotechnical and Computer Science Conference, vol. B, pp. 183–186 (2009)

[15] Pan, J.-I., Yung, C.-J., Liang, C.-C., Lai, L.-F.: An intelligent homecare emergency service system for elder falling. In: World Congress on Medical Physics and Biomedical Engineering 2006, pp. 424–428. Springer, Heidelberg (2006)

[16] Rubenstein, L.Z., Josephson, K.R., Robbins, A.S.: Fall in the Nursing Home. Ann. Intern. Med. 121, 442–451 (1994)

[17] Ubisense (April 23, 2010), http://www.ubisense.net

[18] Zenio, Zenio Fall Detector, http://www.zenio.be/product/8.html

[19] Zhang, T., Wang, J., Liu, J.: Fall detection by wearable sensor and one-class SVM algorithm. Lecture Notes in Control and Information Science, pp. 858–863 (2006)

iRoute: Cognitive Support for Independent Living Using BDI Agent Deliberation

Shabbir Hossain, Kasper Hallenborg, and Yves Demazeau

Abstract. Self-care will be a key driver to save the economy in Western countries, and technology can support it. This paper proposes to use the well-known Belief-Desire-Intention (BDI) [3] [8] architecture, in addition to guidance mobility patterns in order that people with dementia, their caretakers and relatives can feel more safe while maintaining freedom of mobility. Agent architectures are evaluated and a BDI approach taking care of user preference and routes is embedded in an existing prototype that can predict individual walking routes of elders and assist them to get back on track if they start getting lost.

1 Introduction

Dementia is extremely care demanding for both the patient with dementia and the relatives, even though the patients are physically able to handle most daily activities. Therefore it sums to a very high cost of informal, social and medical care, which is estimated to be US$ 604 billion in 2010, and 70% of the costs are spent in Western Europe and Northern America. The term dementia represents a combination of symptoms and among them, memory loss is one of the progressive disorders that affects patient's ability to perform his/her daily activities. Besides, it is hard to accept in the initial phase, when disorders often are impermanent, constrain of the freedom of mobility. Therefore, some solutions should be available to support independent living of patients with Alzheimer's disease, which is a common form of dementia. In outdoor environments Global Positioning System (GPS) trackers and localizers

Shabbir Hossain · Kasper Hallenborg
Maersk McKinney Moller Institute, University of Southern Denmark,
5230 Odense M, Denmark
e-mail: {shho, hallenborg}@mmmi.sdu.dk

Yves Demazeau
Laboratoire dInformatique de Grenoble - CNRS, BP 53 - 38041 Grenoble, France
e-mail: yves.demazeau@imag.fr

J.M. Corchado et al. (Eds.): Trends in PAAMS, AISC 90, pp. 41–50.
springerlink.com

provide the ability to either track down at specific coordinates the current location of the tracker or continuously follow the position by monitoring coordinates transmitted via General Packet Radio Service (GPRS) or Short Message Service (SMS). Such aids are invaluable to increase the safety of both patients and their relatives [10]. When the disorders are becoming more severe, patients often have to move from care homes, where zone detectors and alarm systems warn the staff if patients are leaving the safety zones, and in many cases patients are imprisoned in these locked zones. GPS trackers and localizers solve to some extent the mobility issue, but the overhead of awareness is still burdening either the care staff or relatives, and patients have the feeling of being dependent of care. Therefore we seek to develop an intelligent mobile device that can assist Alzheimer's patients during their travels. The device is developed in the vision of calm technology, which is a notion from ubiquitous computing [15], which encompasses that the system will be silent unless it really needs the attention of the user - we call it the iRoute system. Repeated moving patterns are observed by the system that will try to learn the usual walking routes of the person being observed. Then it will be able to subconsciously monitor if the person deviates from the predicted route and trigger the system, which then shall try to guide the person back on track using oral instructions. In this approach the safety of patients are maintained, as the person can still be localized by staff or relatives, if the person remains lost, but it improves the self-esteem through self-care for the patients. Using mobile phones, the patients will not even feel stigmatized by wearing a special aid, and in a few years mobile phones will become naturally attached to all people throughout their lives and forgetting the device will become less likely than proprietary devices, and route-learning capabilities of mobile phones might serve other assistive purposes in previous life phases, so the iRoute system will already have a strong prediction engine when the person becomes affected by Alzheimer's.

In the next sections, related works that motivate to use the BDI approach to provide cognitive support for the people with Alzheimer's disease of the independent living domain will be followed through a use-case scenario presenting the proposed problem and the solution for the current research work. This is followed by the model of the solution which presents the system overview and will be followed by the implementation that represents the BDI approach using Jadex[1] and an experiment with the system based on the scenario. Finally, discussions on the research work and conclusion will point to further improvements for future work.

2 Related Works

In past years, extensive research has been conducted in outdoor localization systems to track and predict user movement, among them the project called *Opportunity Knocks (OK)* [13]. The project has been designed for navigation guidance to cognitive impaired persons to travel in their community. The system is based

[1] Jadex is an open source project for implementing BDI agents available at: `http://vsis-www.informatik.uni-hamburg.de/projects/jadex/`.

on client-server architecture where the client side consists of a mobile phone with a GPS sensor and the server side consists of a learning and inference engine that incorporates information about user, public transport and geographical data. The inference engine of the system can predict a user's journey information from the user's past activity and can detect and correct route error with transit information in real time. The approach that differentiates the OK system from our work is the agent approach with BDI model. The success of BDI to model human behavior is substantial [12]. The reason behind the success is that BDI core components of the agent framework can translate to the way humans reason about a problem.

There exist other research works [1] [6] [5] on outdoor localization systems but most of them are intended for tourist services. Among them The Wireless Tourist Guide System [7] is an application that uses BDI approach with case-based reasoning (CBR). The work is based on tourist-guide services using Multi Agent System through mobile devices. The system is based on decision process and goal-oriented adaptive agents developed on CBR architecture. The system provides tourist routes with available time and suggests other attractions such as restaurants, museums, monuments and cruises along the route. The tourist guide application is focused on services such as tourist routes for healthy people rather than people with cognitive impairment.

3 Use-Case Scenario: Cognitive Support for a Person with Alzheimer's Disease in His/Her Travel

In order to validate our system, we designed a use-case scenario that can help to show the reasoning approaches of the BDI agent that are used to solve specific problems of people with Alzheimer's disease in their travel. The proposed scenario is designed for a person with Alzheimer's disease who wants to travel from his residence to different places e.g. shops, parks, a friend's house. People with Alzheimer's often forget their routes while they are traveling. They need to be tracked at every stage of their travel otherwise they may get lost from their current route and definitely need more safety than other healthy people.

A user with a desire to travel towards a destination at a certain time has been chosen as our prime scenario for our current research work. Based on the user's previous history, the BDI agent will try to predict a route or several routes of specific destination or destinations. The primary goal of the BDI agent is to track the user, and therefore it is set to the predicted route for a certain destination, but in the case of several possible predicted routes, the BDI agent will try to predict a certain route after user progress on the current route by matching with all other predicted routes. The BDI agent also updates the current position of the user in the web server where the user is represented in a geographical map. Relatives or care staff can keep track of the user's progress on that map continuously. The person with Alzheimer's tends to forget the route and keeps walking off the route. If the user walk off the predicted route,

- In case of a single predicted destination, the BDI agent will help by creating another to guide the user to walk towards the nearest destination that follows the predicted route.
- In case of multiple predicted destinations, the BDI agent will process the guidance in two steps,
  1. firstly by providing route-direction information about all the nearest possible routes that follow the predicted route or routes and
  2. secondly when the user starts progressing on a specific direction then the BDI agent will guide the user till the predicted route or routes.

If the user keeps walking on a route that is not stored in the user's previous history, then the agent will notify the care staff or relatives by sending a warning notification. The BDI agent will keep track the user until the user reaches the destination which is the primary goal of the agent.

4 System Overview

The overall architecture of the iRoute system is shown in Fig. 1. The functionality of the system is composed of three main parts,

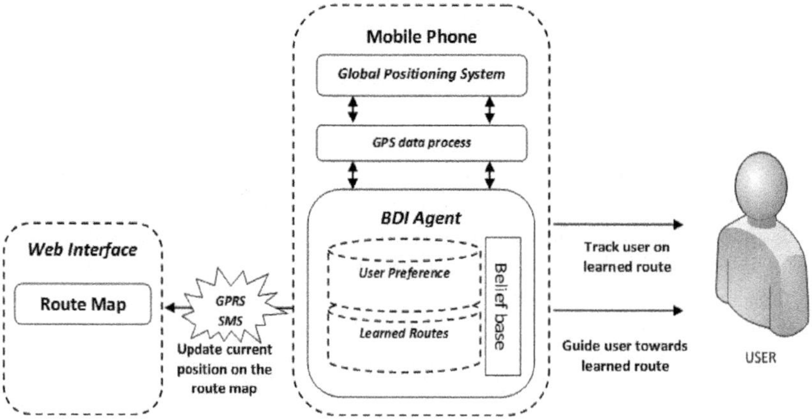

Fig. 1 iRoute system architecture

GPS Data Processing. Generally, the GPS data used for positioning the user may not meet the required precision for using in the system. The result may represent a strong deviation between the received position and the actual position. The processing of the GPS data for error correction and map matching is based on previous research works [11] [9], which are not discussed in this paper. Processed GPS data that are used in the agent's internal reasoning denoted by the function

$f_{gpsdata}(t, pos)$ where t and pos represent time and position. Position pos is represented by a set of GPS coordinate data $pos = \{longitude, latitude, altitude\}$ that is used to locate the user.

Internal Reasoning. Internal Reasoning of the BDI agent using processed GPS data to track and guide user. The reasoning of the BDI agent is separated into two parts,

1. The Route prediction and matching is based on user preferences and learned routes information stored in the BDI agent's belief base. The knowledge about user preferences and learned routes are described in short below,
   a. User preference represents information about an activity. The user preference of an activity denoted by UserPref(a) is the function of activity a such as traveling, taking meals, shopping, activity occurrence occ_a, time T_a of the activity such as starting time for travel activity and location loc of the activity such as destination location for travel activity at time T represented by,
   $$UserPref(a) = f(a, occ_a, T_a, loc_a) \tag{1}$$
   The user preference describes a certain type of user activity such as travel activity that can occur at certain time such as starting time for travel activity at certain location such as destination location for travel activity. The user preference can have preference level of an activity based on the activity occurrences. The activity occurrences categorized by occurrences happening on a daily or weekly or monthly are represented by $occ_a = \{daily_a^{1..n}, weekly_a^{1..n}, monthly_a^{1..n}\}$ with minimal cardinality 1.
   b. Routes that are stored from the user's previous travel history are known as learned routes. Routes are stored as a set of positions from the current location till the destination location along with time intervals. Learned routes are aggregated and filtered information from all previous travel history of the user. Routes are presented by Route(loc_a) of location of certain activity a such as for travel activity the location represent the destination location of travel activity, is defined by the function of time t, position pos and location of activity loc_a,
   $$Route(loc_a) = f(t, pos, loc_a) \tag{2}$$
   The position represented by a set of stored GPS coordinate data $pos = \{longitude, latitude, altitude\}$.
2. The agent tracks the user towards the predicted route or routes until the user reaches one of the predicted destinations, but he/she can have more than one predicted route at a certain time of his journey. During the travel,
   a. if the user walks off the route, then the agent tries to guide the user using voice-activated navigation to walk back on the last predicted route. If the user avoids voice-activated navigation and keeps moving in the wrong path, then the agent will notify either the care staff or the relatives.

b. if the user forgets the direction to walk towards his desired destination from a multiple route point, then the agent will provide all the information needed including direction towards all destinations from that point.

Localizing User. Localize the user, by processing GPS data and updating the latest position of the user in the web interface for care staff and relatives. This location update is visible to care staff, relatives and relatively increases the safety for the user, in case of irrational behavior related to dementia, but attention is not required until the usual travel has been exceeded.

5 BDI Implementation Using Jadex

The Jadex agent framework provides an FIPA [2] compliant agent environment and a set of development tools for deploying Multi-Agent Systems (MAS) based on the notion of BDI. The core component of the Jadex agent is based on beliefs, goals and plans.

- *Beliefs* represent information that allows a Java object to be stored as a fact or set of facts. Beliefs can be updated, removed and queried by the agent. Beliefs are used as preconditions of plans and creation condition or target condition or as context condition of goals [4]. Any change of a belief that can create a new goal or drop a goal or terminate a plan, will turn the situation into an invalid plan.
- *Goals* represent the desire that the agent wants to achieve rather than how to achieve that desire. Goal condition is used to realize if the goal is achieved or not. In Jadex, a goal life cycle has been used to differentiate between active goals, option goals and suspended goals. A goal deliberation process is used for evolution of the goal states. Based on goal life cycle, different types of goals i.e. perform goal, achieve goal, query goal, maintain goal have been introduced in Jadex [4].
- *Plans* are used to achieve the goal by executing a set of or sets of actions. In Jadex, the plans consist of two parts: head and a consequent body. The plan head is stated in the Agent Definition File(ADF) that defines in which condition the plan body will be executed and the plan body is implemented in a pure Java class that includes handling sub goals, events and incoming FIPA compliant Agent Communication Language (ACL) messages [4].

Rather than plans, an XML based Agent Definition File (ADF) is used to specify the general structure of beliefs, goals and plans head. The execution model of Jadex separates the processing of goal and execution of plan [14]. The execution model of Jadex processes the hierarchy of goal and plans which can start from top-level goal or goals to plans and plans to sub-goals and so on. The example of processing of goal and sub-goal over plans of our implementation is described in Fig. 2.

[2] FIPA Foundation for Intelligent Physical Agents, 1996.`http://www.fipa.org/`.

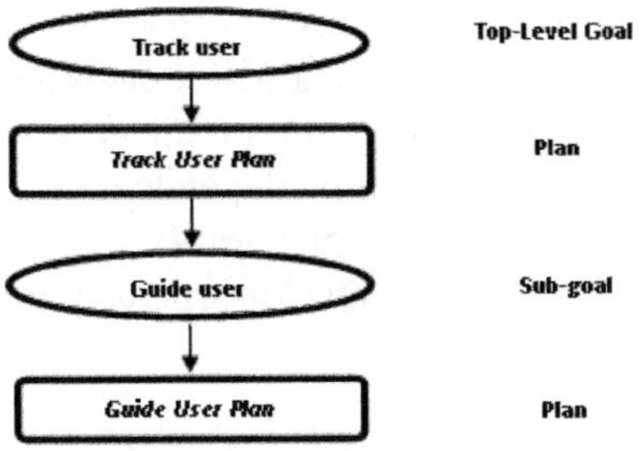

Fig. 2 Goal deliberation of the BDI agent in iRoute system

The reasoning part of the BDI agent is the core part of our research work. Therefore, we have discussed the reasoning part of the BDI agent in later sub sections.

5.1 Prediction of the User's Destination

The prediction of the user's destination is based on the time of the activity, location of the activity and occurrence of the activity from the user preference described in equation 1. The prediction of the user's destination is processed in two phases:

- In the first phase, predicted locations are populated if the time range $t_r = \{t_1, t2, ..., t_c, ..., t_n - 1, t_n\}$ generated by adding and deducting specific time variable with the current time t_c do exist inside the set of time T_{travel} that belongs to $UserPref(a_{travel})$ where $a_{travel} = alltravelactivity$ defined at equation 1. The sum of predicted location $\sum_{i=1}^{n} f_i(loc_{travel})$ is defined by sum of function $f_i(loc_{travel})$,

$$f_i(loc_a) = \{\exists t_r \in T_{travel} \implies loc_{travrl} \mid (T, loc_{travel}) \in UserPref(a_{travel})\}$$

- The resulted locations from first phase then filtered by the activity occurrences occ_a defined in user preference of a certain activity at equation 1 which gives optimized result of predicted location by prioritizing the results. The prioritization rule put more weight in the function of predicted location that occurred daily than weekly or monthly and same for weekly than monthly.

5.2 Reasoning for a Single Predicted Destination

Fig. 2 shows that the BDI agent is initialized with the primary goal *track user*, which includes a *target condition* that checks in the *belief* for information about whether the user reaches the predicted destination or not. The goal *track user* is triggered by the plan *track user plan*. The implementation of the plan body, which consists of actions such as checking the distance between the current position of the user received from the GPS and the predicted destination, matches the user position with the learned routes that lead to the user's predicted position. If the current position of the user matches with the learned routes, then it is considered that the user is walking on the predicted learned route otherwise the user is off the predicted learned route. If the user is off the predicted route, then the agent will create a sub-goal *guide user*, which will be dependent on top-level goal *track user*. The goal *guide user* is triggered by the plan *guide user plan* where the user is guided by oral instruction towards the nearest known position that follows the predicted route.

5.3 Reasoning for Multiple Predicted Destinations

As we mentioned before, multiple destinations can be predicted by the BDI agent at a certain time of the travel. In a situation with multiple destinations, checking the current position with the predicted multiple routes to optimize the prediction about the user's destination takes place while the user is progressing on the route. During the process of optimizing the prediction of the user's destination,

- if the user walks off the predicted routes before the completion of the optimization process of the predicted destination of the user, the *guide user plan* is processed into two phases, firstly by providing oral information of walking directions about all possible learned routes from the user's current position by creating another sub-goal *directional info guide*, so that the user can decide the correct direction. The success of the goal will be reached when the user will start walking on a certain direction. Secondly, the user will be guided while progressing on a certain direction by creating another sub goal *guide user to known route* when goal *directional info guide* is succeeded.
- if the user is approaching towards a route point from where the possibility of facing difficulties deciding about the correct direction of the user's desired destination, the reasoning part of the agent in *track user plan* will provide oral instruction regarding all possible directions from that point.

6 Experiment and Results

In order to validate our system, we have chosen a participant with mild cognitive impairment and tested the system based on the described scenario in section 4. The results shown in Fig. 3 are taken from the web interface. Fig. 3a shows, the result of single predicted destination (from A to B) experiment and Fig. 3b shows the result of multiple predicted destinations (from A to B or A to C or A to D) experiment.

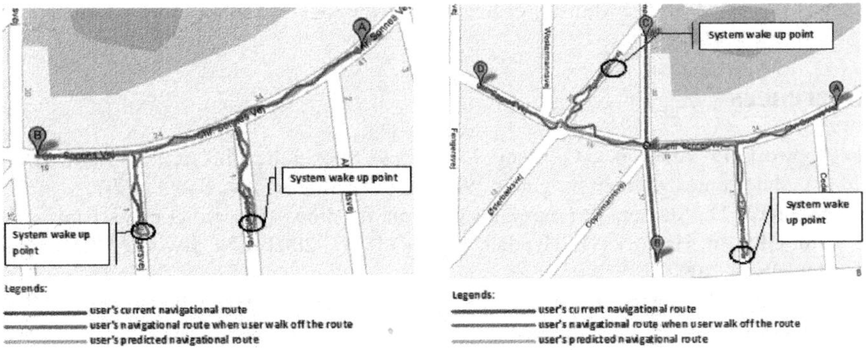

(a) single predicted destination (b) multiple predicted destination

Fig. 3 experiment results for iRoute system

From the figures above, we can observe that the agent reacts when the participant deviates from the predicted route and guides the participant towards the predicted route in case of both the single predicted destination and the multiple predicted destination.

7 Discussion and Conclusion

The work in this paper represents a preliminary approach to provide cognitive support for people with Alzheimer's disease in their travel using an adapted BDI agent. The work here was focused on specific problems and solution described in the use-case scenario. We have described an approach to use the natural mapping between the BDI architecture and human cognition to create a cognitive support tool that can become a natural aid to demented persons in the spirit of calm technology. The proposed system can use ordinary mobile phones, so that people do not feel stigmatized by wearing dedicated aids. The improvement of the system includes route blockade, change of user decision regarding destination. Daniel Ashbrook and Thad Starner [2], have done an extensive work on user movement prediction with the Markov Model. The use of the Markov model in our prediction model can be used as a part of improvisation of iRoute system. Context aware services like weather information during the travel and user's expected environment can be provided by adding the environmental information in the belief base of the BDI agent that can increase the functionality of our system. The improvement of the system focus on the integration of enhanced learning mechanism of the BDI agent that can predict situations such as route blockade, change of decision of the user regarding destination from user behavior, integration of user input and use of visual output such as text and map to provide more correct guidance.

Acknowledgements. The presented work has been conducted under the IntelliCare project, which is supported by The Ministry of Science, Technology and Innovation in Denmark.

References

1. Abowd, G.D., Atkeson, C.G., Hong, J., Long, S., Kooper, R., Pinkerton, M.: Cyberguide: A mobile context-aware tour guide. Wireless Networks 3(5), 421–433 (1997)
2. Ashbrook, D., Starner, T.: Learning significant locations and predicting user movement with GPS. In: Horrocks, I., Hendler, J. (eds.) ISWC 2002. LNCS, vol. 2342, Springer, Heidelberg (2002)
3. Bratman, M.: Intention, plans, and practical reason. Harvard University Press, Cambridge (1987)
4. Braubach, L., Pokahr, A., Lamersdorf, W.: Jadex: A BDI agent system combining middleware and reasoning. In: Klusch, M., Unland, R., Calisti, M. (eds.) Software Agent-Based Applications, Platforms and Development Kits (2005)
5. Burrell, J., Gay, G., Kubo, K., Farina, N.: Context-aware computing: A test case. In: Borriello, G., Holmquist, L.E. (eds.) UbiComp 2002. LNCS, vol. 2498, pp. 647–653. Springer, Heidelberg (2002)
6. Cheverst, K., Davies, N., Mitchell, K., Friday, A., Efstratiou, C.: Developing a context-aware electronic tourist guide: some issues and experiences. In: Proceedings of the SIGCHI Conference on Human Factors in Computing Systems, pp. 17–24. ACM, New York (2000)
7. Corchado, J.M., Pavón, J., Corchado, E., Castillo, L.F.: Development of CBR-BDI agents: A tourist guide application. In: Funk, P., González Calero, P.A. (eds.) ECCBR 2004. LNCS (LNAI), vol. 3155, pp. 547–559. Springer, Heidelberg (2004)
8. Dennett, D.C.: The Intentional Stance. MIT Press, Cambridge (1987)
9. Greenfeld, J.S.: Matching GPS observations to locations on a digital map. In: 81th Annual Meeting of the Transportation Research Board (2002)
10. Jensen, L., Mänsson, I., Holthe, T., Hurnastid, T., Gušnadóttire, Ž.: How cognitive assistive technology can be useful for and a benefit to persons with dementia in their daily living. guidelines for service delivery of cognitive assitive technology to persons with dementia. In: Challenges for assistive technology: AAATE 2007, pp. 227–232 (2007)
11. Martin, J.D., Boll, S.: Dynamic GPS-position correction for mobile pedestrian navigation and orientation. Oldenburg (2005)
12. Norling, E.: Folk psychology for human modelling: Extending the BDI paradigm. In: Proceedings of the Third International Joint Conference on Autonomous Agents and Multiagent Systems, vol. 1, pp. 202–209. IEEE Computer Society, Los Alamitos (2004)
13. Patterson, D.J., Liao, L., Gajos, K., Collier, M., Livic, N., Olson, K., Wang, S., Fox, D., Kautz, H.: Opportunity knocks: A system to provide cognitive assistance with transportation services. In: Davies, N., Mynatt, E.D., Siio, I. (eds.) UbiComp 2004. LNCS, vol. 3205, pp. 433–450. Springer, Heidelberg (2004)
14. Piunti, M., Ricci, A., Braubach, L., Pokahr, A.: Goal-directed interactions in artifact-based mas: Jadex agents playing in cartago environments. In: IEEE/WIC/ACM International Conference on Web Intelligence and Intelligent Agent Technology, WI-IAT 2008, vol. 2, pp. 207–213. IEEE, Los Alamitos (2009)
15. Weiser, M., Brown, J.S.: The Coming Age of Calm Technology. Springer, New York (1997)

Dynamic Distribution of Tasks in Health-Care Scenarios

Carolina Zato, Ana de Luis, and Juan F. De Paz

Abstract. This paper presents a multiagent system that use an autonomous deliberative case-based reasoningagent to design an efficient working day. The system has been developed to plan and distribute tasks in a health care scenario, specifically in geriatric residences. This model generates a planning of tasks, minimizing the resources necessary for its accomplishment and obtaining the maximum benefits.For this purpose, the queuing theory and genetic algorithms have been include in a CBRarchitecture to obtain an efficient distribution. To evaluate the model, the obtained results have been compared with a previous method of planning based on neural networks.

Keywords: multiagent systems, queuing theory, genetic algorithm, task scheduling, health-care.

1 Introduction

During the last three decades the number of Europeans over 60 years old has risen by about 50%. Today they represent more than 25% of the population and it is estimated that in 20 years this percentage will rise to one third of the population, meaning 100 millions of citizens [1]. This situation is not exclusive to Europe, since studies in other parts of the world show similar tendencies. In the United States of America, people over 65 years old are the fastest growing segment of the population and it is expected that in 2020 they will represent about 1 of 6 citizens totalling 69 million by 2030. Furthermore, over 20% of people over 85 years old have a limited capacity for independent living, requiring continuous monitoring and daily care, for this reason it is important to create a task planner and control system for elderly people.

Carolina Zato · Ana de Luis · Juan F. De Paz
Department of Computer Science and Automation, University of Salamanca
Plaza de la Merced, s/n, 37008, Salamanca, Spain
e-mail: {carol_zato,adeluis,fcofds}@usal.es

J.M. Corchado et al. (Eds.): Trends in PAAMS, AISC 90, pp. 51–58.
springerlink.com

Nowadays, a great amount of dynamicproblems can be considered as planning problems. In this kind of problems the aim is to obtain a plan, as well as obtaining a scheduling of the actions along the time, assigning the necessary resources and taking into account restrictions like the benefits. Practical planning has been applied in many different fields and has been combined with different artificial intelligence techniques [2] [10]. One of the main challenges has been the development of intelligent systems focused on the improvement of health care sciences that can be applied in hospitals, residences, etc. [3].

The system presented in this paper includes deliberative agents using a case-based reasoning (CBR) architecture, as a way to implement adaptive systems to improve assistance and health care support for elderly and people with disabilities. Agents in this context must be able to respond to events, take the initiative according to their goals, communicate with other agents, interact with users, and make use of past experiences to find the best plans to achieve goals.The planning model is performed by applying CBR systems [5]. The CBR system integrates into various stages of reasoning, techniques to estimate resources based on queuing theory and planning using genetic algorithms.

The planner has been included into the ALZ-MAS multi-agent system [4][7]. ALZ-MAS generates the distribution of tasks among the personal taking into account their abilities and allowing the dynamic distribution of tasks according to their profiles.ALZ-MAS facilitates carrying out tasks in a ubiquitous way, such as plans generation for nurse's and doctors working days, or patient and task assignation. These tasks are executed in a distributed way inside the intelligent environment and are independent of the location of the processing devices due to the use of mobile devices and wireless networks. The previous planner of ALZ-MAS calculated the distribution of tasks using neural networks to establish the final sequence of actions. Now, this model has been improved using the queuing theory and genetic algorithms.

This article is divided as follows: section two presents the proposed model and section three describes the results obtained and the conclusions.

2 Proposed Model

The model proposed in this paper focuses on developing a planning mechanism to coordinate the agents. The roles of the agents are:

1. **Patient role.** It manages the patient's personal data and behaviour (monitoring, location, daily tasks, and anomalies). Every hour validates the patient location, monitors the patient state and sends a copy of its memory base to the manager agent in order to maintain backups.
2. **Personnel role.** It represents the different workers. This role can have different specializations according to the profile of the workers and his capabilities. In this case, the specialized profiles are doctors, nurses and nurse assistants.
3. **Planner role.** Design the overall plan to be implemented by the organization. Sets the number of personnel agents and makes the distribution of tasks depending on the specialized role they play. Replan depending on the size of the input queue or inability to accomplishwith a plan.

4. **Distributor role.** Distributing tasks according to its completion by the agents and checks that each task is being processed within time limits to serve the plan.
5. **Manager role.** This agent controls the security (patients location, locks, alarms, etc.) and manages all the information of the system (record database, incidences, etc.).

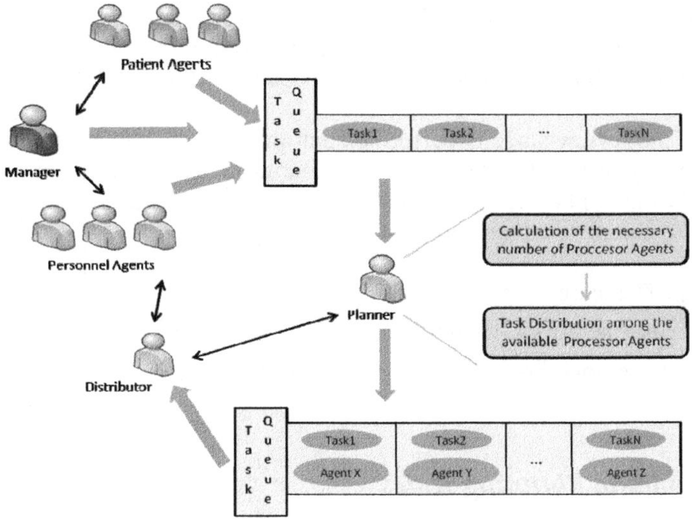

Fig. 1 Outline of proposed model

In the figure 1 the different agents of the system and some of the interactions among them are represented. In the upper corner of the figure appears the task list that store the activities to carry out in the multi-agent system and in the centre of the image, the planner agent that generates the planning.

To carry out the planning process, the agent follows the planning model CBR-BDI. The first point in the definition of a CBR-BDI model is the definition of case (1).

$$C = \{t_i / t_i = (ids, idt, l, f, p, d, a), i = 1...n\} \quad (1)$$

Where t_i represents the task *i*, *ids* is the identifier of the task type, *idt*the process, *l*localization (x, y, floor), *f* date of entry, *p* deadline date, *d* duration, *a* the ID of the agent that performed the task. The application of the different stages of reasoning is performed as follows:

Recovery. During this stage the most similar cases to the current case cn+1 are retrieved. The most similar cases are those containing more tasks of the same type as the tasks that are currently queued in the system. The number of cases retrieved is predefined to subsequently implement the adaptation phase. The set of recovered cases are called $C_r \subseteq C$.

Adaptation. During this stage, the recovered information Cr is adapted from cases of memory to generate the entire plan corresponding to the case cn+1. The information recovered is adapted by applying queuing theory and genetic algorithms. On one hand, the recovered information is used to determine the arrival rate and service for each of the tasks and thus using queuing theory we can determine the number of agents needed to run the indicated tasks. Recovered cases are the basis for constructing initial chromosomes in genetic algorithms.

Review. The review stage is performed automatically as the agents are finalizing tasks. The agent updates the duration of the tasks as they are completed and in turn, makes new plans if any notice is received from the processor agents on the inability to complete a plan under time constraints provided.

Learning. The learning phase is limited to store the case when the day is finished. The new case memory C 'is defined as follows: $C' = C \cup c_{n+1}$ To limit the size of cases of memory, some are removed from memory if they exceeded a predefined age.

2.1 Dynamic Planning Roles

The number of agents that should be available in the system is estimated dynamically. It is intended that the number of agents suits demand to ensure that the system utilization factor ρ is less than 1. This estimate will be done through the use of queuing theory in a model M/G/s, where the arrival rate follows a Poisson distribution (the most commonly used in similar work), the exponential service and the existence of multiple servers (agents).

The problem of planning multiple tasks can be reduced to the case of planning for a single task of each type. Thus, for each task a planning is performed independently so as to calculate the average waiting time and average queue length independently. The average waiting time and the overall average length is reduced to calculating the average values calculated for each of the tasks. In the case of the M/G/s model where s = 1,2,3, ... is the number of agents and given an arrival rate $\lambda_n = \lambda$ = cte, the service rate when there are n processes is defined by the following equation (2) [8] [5].

$$\mu_n = \begin{cases} n\mu & n = 1, 2, \ldots, s-1 \\ s\mu & n \geq s \end{cases} \tag{2}$$

Where μ represents the average service rate for s available agents. This value depends on both the agents and the machine found.

Assuming that the system is in a stable condition, i.e. it meets the utilization factor $\rho = (\lambda / \mu s) < 1$, the queuing theory [8] allows to calculate the probability that n tasks exists (P_n) in the system, the number of tasks (L) in the system.
To determine the optimal number of agents,an estimation that minimizes the cost function is calculated and it depends on the number of agents used and on the waiting time in the queue. The function is defined in a particular way for each service depending on the actual costs of each agent in the system, though the following benefit function is provided (4).

$$f(L, P_0,.., P_{s-1}, \mu', \overline{p}, \overline{b}) = f_b(L, \mu', \overline{p}, \overline{b}) - k \cdot s \tag{3}$$

$$f_b(L, \mu', \overline{p}, \overline{b}) = \begin{cases} (\overline{p} / u')\overline{b} \cdot s \cdot (1-\rho) & \text{si } L \cdot u' > \overline{p} \cdot s \cdot (1-\rho) \\ L\overline{b} & \text{si } L \cdot u' \leq \overline{p} \cdot s \cdot (1-\rho) \end{cases} \tag{4}$$

Where k is a constant associated with the cost of having an agent working, $\overline{b}$ the average benefit of performing the task μ' is the average time to complete the task obtained from the service rate, $\overline{p}$ the average time to execute a task. If the conditions of stability are overpass, f_b is counted only up to the utilization factor 1. The utilization factor ρ varies according to the new services added to the queue till it reaches the utilization factor of 1.

Following the cost function given in (4), a global cost function is introduced (5) that takes into account the implementation of the various services.

$$f(f_1,..., f_k) = \sum_{j=1}^{k} f_i \tag{5}$$

Where f_i is calculated from equation (4). Because sometimes you might not be given the stability conditions, it is necessary to calculate the terms in order of benefit depending on the type of the task so that when you reach the utilization factor of 100%, the process ends calculating the summation terms. Once the optimization function defined in (5) the maximum value is calculated iteratively starting with number of agents equal to 1, the fixed value is the first local maximum that corresponds to the global maximum.

2.2 Task Assignment

Once the number of starting agents is considered to minimize costs, an allocation of tasks between the available agents has to be done. The distribution of tasks among agents is performed so as to ensure as far as possible that it can perform assigned tasks in case of delays or the time to perform a task increases. It is performed so as to maximize the following function (6):

$$\min \sum_{i=1}^{k} \pm \log|1 + x_i - \overline{x}| \qquad (6)$$

Where $\log|1+x_i - \overline{x}|$ takes the sign of the expression $x_i - \overline{x}$, $x_i = t_i - a_{i-1} - c_i$ with t_i the maximum time for completion of the task i, a_{i-1} the accumulative time to perform the tasks i-1 above and finally c_i the time to run the task *i*plus the time of movement from the task *i-1*calculated by Floyd algorithm [11], which is customized according to the agent selected and calculated from the average value of previously executed tasks. Minimizing the differences get all the tasks to have a uniform distribution of the remaining time so it gets easier to achieve them.

The chromosome encoding is performed so that each gene is composed of the elements listed by t_i identified in (1). The crossover operator is defined similarly to the multi-junction used in other problems such as TSP. The operator is defined as follows:

- Select a partial route
- Exchange of direct segments of tasks where ID matches
- The exchanges define a series of matches that relates each of the genes of a chromosome with which occupies the same position in the other parent

Mutation operators define various modes that will be executed randomly, and just those mutations that improve the aptitude of chromosomes will be selected. The defined mutation operators are: exchange order of tasks, exchange of assigning contiguous tasks and changing the allocation of a task.

Elitism operator is defined to keep the percentage of efficient solutions in every generation of population and population size as a constant this involves the replacement of parents by the children chromosomes in generations with the exception to remain with elite chromosomes. The roulette selection is the criteria chosen for this.

The initiation of chromosomes is based on the received tasks; each chromosome is initiated with the tasks of the new case. For each task of the chromosome a sequential search is made in the case and assigns the task to an existing agent so that this association is maintained for the remaining tasks.

3 Results and Conclusions

The previous planning model of ALZ-MAS calculated the distribution of tasks using neural networks to establish the final sequence of actions. This model was implemented by means of a RPTW (Routing Problems with Time Windows) neural network[4],which allows reaching a solution much more rapidly.

In order to compare both systems, we measure the number of replannings carried out by the planner agent. The first planner of ALZ-MAS was applied in a working day and the process was repeated the next day using this new proposed planner. This process was repeated during 5 iterations. The image 1 shows the numbers of replannings carried out for each of the 5 nurses. The y axis represents the number of replannings and the x axis the nurses. The thin lines represent the RPTWplanner of ALZ-MAS and the fat lines the new planner. The colour represents the nurses. The number of replannings is reduced because the new planner creates the plans bearing in mind the intervals among the tasks in the final plan.

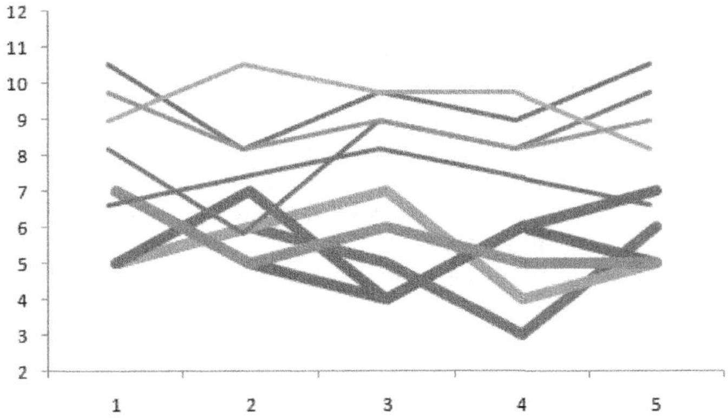

Fig. 1 Numbers of replannings carried out for each of the 5 nurses. Fat lines new planer and thin lines RPTW.

To complete the study the Mann-Whitney U-test [9] was applied. It was a non-parametric test and it is not necessary to make assumptions on the data distribution, as in the t-test. The test determines two values: H0 and H1. H0 shows if the data in both groups presents the same distribution, whereas H1 determines if there is difference in number of replannings. The number of replannings for each worker was compared in both planners. The results establish that the number of replannings comes from different statistical distributions for each workerso, we can conclude that the differences are relevant.

The proposed planner allows reducing the number of replannings because it establishes homogeneous intervals among the tasks and calculates the arrival time of tasks using the previous information. Besides, the planner makes predictions of the necessary personal in order to accomplish all the activities.

Acknowledgments. This study has been supported by the Spanish Ministry of Science and Innovation project TRA2009_0096.

References

1. Camarinha-Matos, L., Afsarmanesh, H.: Design of a virtual community infrastructure for elderly care. In: PRO-VE 2002 – 3rd IFIP Working Conference on Infrastructures for Virtual Enterprises, Sesimbra, Portugal (2006)

2. Zülch, G., Rottinger, S., Vollstedt, T.: A simulation approach for planning and reassigning of personnel in manufacturing. International Journal of Production Economics 90, 265–277 (2004)
3. Foster, D., McGregor, C., El-Masri, S.: A Survey of Agent-Based Intelligent Decision Support Systems to Support Clinical Management and Research. In: Proceedings of the First International Workshop on Multi-Agent Systems for Medicine, Computational Biology and Bioinformatics, pp. 16–35. Utretch, Netherlands (2005)
4. Bajo, J., De Paz, J.F., De Paz, Y., Corchado, J.M.: Integrating case-based planning and RPTW neural networks to construct an intelligent environment for health care. Journal Expert Systems Applications 36(3), 5844–5858 (2009)
5. Martín, Q., Santos, M.T., de Paz, Y.: Investigacion Operativa: Problemas y Ejercicios. Pearson Educacion, London (2005)
6. Kolodner, J.: Case-Based Reasoning. Morgan Kaufmann, San Francisco (1993)
7. Corchado, J., Tapia, D., Bravo, J., García, Ó., Saavedra, A., Alonso, R., García, I.: ALZ-MAS 2.0; A Distributed Approach for Alzheimer Health Care. In: 3rd Symposium of Ubiquitous Computing and Ambient Intelligence 2008. AISC, vol. 51, pp. 76–85. Springer, Heidelberg (2008)
8. Martín, Q.: Investigación Operativa. Prentice-Hall, Englewood Cliffs (2003)
9. Yue, S., Wang, C.: The influence of serial correlation on the Mann–Whitney test for detecting a shift in median. Advances in Water Resources 25(3), 325–333 (2002)
10. Yu, J., Buyya, R.: Scheduling scientific workflow applications with deadline and budget constraints using genetic algorithms. Scientific Programming 14(3), 217–230 (2006)
11. Hougardy, S.: The Floyd–Warshall algorithm on graphs with negative cycles. Information Processing Letters 110(8-9), 279–281 (2010)

Automating the Development of Agent-Based Warehouse Control Systems

Jacques Verriet, Bruno van Wijngaarden,
Eef van Heusden, and Roelof Hamberg

Abstract. Warehouses play a critical role in supply chains: they are responsible for the distribution of goods of many suppliers to many customers. Because of increasing customer demands, warehouse processes are becoming too complex to be controlled optimally by their traditional centralised control systems. Several results in the literature show that decentralised controllers are a viable alternative. Unfortunately, the development of effective decentralised warehouse controllers is quite complex, as one has to design the desired system behaviour from the behaviours of its components. This paper proposes a generic architecture that supports the design of agent-based warehouse controllers. This architecture is based on two types of generic components having behaviours with business rule plug-ins. This paper describes the new architecture and how it can be used to automate the development of warehouse control systems.

Keywords: Warehouse control systems, agent-based control, JADE, automatic code generation.

1 Introduction

Although they are not very visible to the general public, warehouses play an important role in modern society; they enable the efficient distribution of goods from many suppliers to many customers. They achieve this by receiving goods from suppliers, providing storage for them, and shipping them when they are ordered. To perform these operations, a warehouse generally has to prepare the goods.

Jacques Verriet · Roelof Hamberg
Embedded Systems Institute, P.O. Box 513, 5600 MB Eindhoven, the Netherlands
e-mail: {jacques.verriet,roelof.hamberg}@esi.nl

Bruno van Wijngaarden · Eef van Heusden
Vanderlande Industries, Vanderlandelaan 2, 5466 RB Veghel, the Netherlands
e-mail: {bruno.van.wijngaarden,eef.van.heusden}@vanderlande.com

J.M. Corchado et al. (Eds.): Trends in PAAMS, AISC 90, pp. 59–66.
springerlink.com

Before goods can be stored, they may need to be repacked in special storage containers, which can be handled in an automated manner. Once goods are ordered, they need to be shipped to the warehouse's customers. The preparation for shipping generally involves repacking of goods into special delivery containers. De Koster et al. [2] and Rouwenhorst et al. [8] provide elaborate overviews of common warehouse processes and techniques to optimise them.

It is not uncommon for a modern warehouse to deliver goods to different types of customers. These may include other warehouses, various types of stores, and internet customers. Each type of customer has its own delivery requirements. For instance, a warehouse can save its supermarket customers a lot of money, if it delivers goods in such a way that the supermarkets' shelves can be replenished efficiently. This value-added service, called store-friendly delivery, requires that similar supermarket products are packed in the same delivery container and that these containers are arranged in a specific sequence. This packing and sequencing greatly reduces the travel time of the supermarkets' replenishment personnel, but requires a complex flow of goods in a warehouse.

Warehouse control. The planning and scheduling of warehouse operations are controlled by a warehouse control system. Traditional warehouse control systems are built around a centralised database containing full knowledge of the warehouse status. These warehouse control systems allow a high warehouse performance, because all optimisation information is available in one place.

It is already quite a challenge to achieve a good warehouse performance under normal operating circumstances. In case of regularly occurring exceptions, this becomes practically impossible. It is especially challenging for centralised warehouse control systems, which need to have a detailed answer to a wide variety of exceptions. The full system knowledge hampers the scalability of centralised warehouse control systems, because all event handling involves access to the central database, making it a potential bottleneck for complex warehouses. Moreover, it is often difficult to reuse warehouse control functionality for different warehouses or to exchange similar functionalities.

Agent-based control. Multi-agent systems overcome the limitations of traditional warehouse control systems. Agent-based warehouse control systems consist of a collection of cooperating autonomous agents, each controlling the operations in a limited part of a warehouse. The modularity of agent-based control systems supports maintainability and reusability. The autonomy of the agents enables agent-based control systems to adapt to changing circumstances.

Kim et al. [6] study agent-based control for a warehouse with a man-to-goods picking concept and Graves et al. [3] for one with a goods-to-man concept. They both propose a hierarchical agent-based warehouse control system, in which the high-level agents are responsible for providing an initial schedule. The low-level agents start with such a schedule, but constantly try to find a local schedule that best fits the real-time conditions. If they find such a schedule, they negotiate with the high-level agents for a schedule update. The warehouse-specific results of Kim et al. [6] and Graves et al. [3] show that their constant replanning has a positive influence on the system performance, especially in case of exceptions.

Ito and Abadi [4] have a more generic approach to agent-based warehouse control; they propose a layered architecture for agent-based warehouse control. Their architecture consists of three main subsystems: communication, inventory planning, and material handling. Each subsystem is implemented by a collection of agents each with its specific responsibilities.

Hiel et al. [5] address the challenge of developing agent-based warehouse control systems. They use agent organisations to identify agents and their connections. They use roles enacted by agents or agent organisations to implement planning and scheduling functionality. Moneva et al. [7] also use roles in their warehouse control framework. They standardise agent roles and agent interactions to reduce the effort of developing agent-based warehouse control systems.

Outline. In this paper, we go one step further in reducing the warehouse development effort. We present a generic architecture that standardises not just the agents and their interfaces, but also the agent functionalities. This is achieved by using generic behaviours with local business rule plug-ins, which can be used to make a behaviour application specific. This standardisation allows warehouse control systems to be generated automatically by identifying the agents, their interconnections and behaviours, and by specifying the behaviours' business rules.

The paper is organised as follows. Section 2 explains the components and behaviours of the new warehouse control architecture. The implementation of the generic architecture is described in Section 3. In Section 4, we present the initial results of the validation of the new architecture. Section 5 provides the current status of the architecture and an outlook into the future.

2 Generic Warehouse Control Architecture

This section describes the generic warehouse control architecture, which will form the basis for automatic generation of warehouse control systems.

Components. The main elements of the architecture are two types of generic components: *stock planners* and *device managers*. The main responsibility of a stock planner is the delivery of goods to its (local) customers, which requires it to maintain a level of local stock. The device managers provide the interface between the (abstract) stock planners and the warehouse equipment. They represent the plant and are responsible for scheduling the work assigned by the stock planners.

Like all successful applications of agent-based warehouse control in the literature, our warehouse control architecture has a hierarchical structure. The warehouse components comprise a system with two layers: a planning layer with stock planners and a scheduling layer with device managers. Together the stock planners form a tree. This tree structure resembles the hierarchical warehouse structure identified by Moneva et al. [7]. Each stock planner has a number of device managers that couple it to the warehouse equipment.

We will use the system shown in Figure 1 as a running example. It shows a case picking *module* with a *palletiser* and three *case picking cells*. Each cell consists of two *pick fronts*, a *reserve*, two *case pickers*, and a *tray miniload*. Figure 2

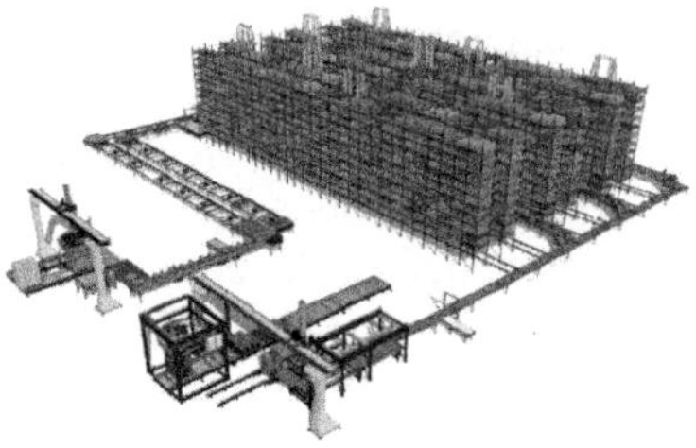

Fig. 1 Case picking module

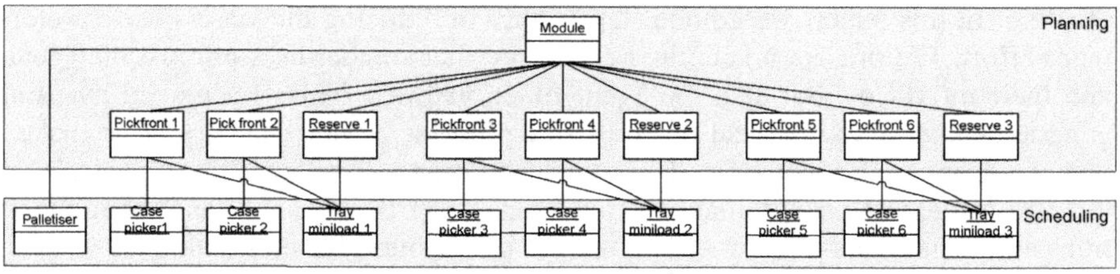

Fig. 2 Case picking module components

shows the corresponding component structure. The module, pick fronts and reserves are stock planners; the palletiser, case pickers and tray miniloads are device managers. If the module is asked to deliver a pallet, it asks its children, the pick fronts and the reserves, for the corresponding cases. The pick fronts' stock consists of trays, from which the associated case picker can pick cases. The reserves also store trays of cases, but they can only be used to deliver entire trays. If a pick front cannot deliver a certain product, it can be replenished from a reserve. The corresponding tray miniload can pick entire trays and store them in the pick fronts' stock.

Figure 2 shows that there need not be a one-to-one relationship between stock planners and device managers. The tray miniloads are associated to two pick fronts and one reserve. It also shows that device managers can be connected. The case pickers of one case picking cell are peers. This means that they can take over each other's work in case one of them breaks down or is in maintenance.

Behaviours. The main challenge in developing an effective decentralised warehouse control system is designing its components' behaviours that together provide the desired system behaviour. Existing work on agent-based warehouse control provides little support for this. We have standardised component behaviour to alleviate the system design effort. For warehouse operations, we have defined a collection of generic behaviours. These behaviours are triggered by the reception of messages matching standardised interfaces. In this section, we will describe some of our architecture's behaviours and the corresponding messages.

Delivery Behaviour. The main goal of the stock planners is the delivery of goods to its customer stock planners. Several behaviours have been defined to control this delivery. They have been specified as UML activity diagrams. An example is the *assign work* behaviour in Figure 3. This behaviour is triggered by the reception of an *assign work* message; this is a standardised interface specifying the goods to be delivered and the container type in which to deliver them. After receiving the message, the assign work behaviour sends *stock inquiry* messages to its child stock planners to ask for the availability of the requested goods. It then waits for the corresponding *stock reply* messages and computes the available stock level. If the stock level does not suffice, the behaviour issues replenishment work by sending an *assign replenishment* message to its stock planner agent. This message will be handled by the stock planner's *assign replenishment* behaviour, which replies with an *accept replenishment* message. The assign work behaviour ends by sending an *accept work* message to acknowledge the original assign work message.

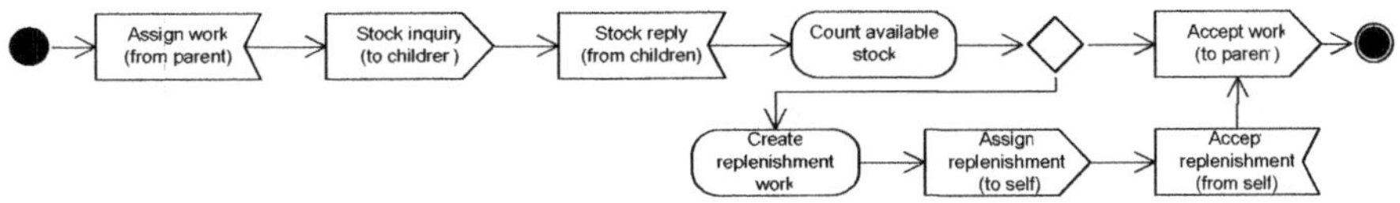

Fig. 3 Assign work behaviour

Replenishment Behaviour. The description of the assign work behaviour showed the existence of replenishment assignments. Because these are triggered by the orders being issued, they may take a long time to be fulfilled. This may not be acceptable for popular products; these have to be delivered so often, that they should be in stock at all times. This is the responsibility of the *min-max replenishment* behaviour. This behaviour gets triggered when goods have been delivered by a stock planner. The behaviour will then check whether the stock level for a certain product has dropped below a minimum. If so, the behaviour will issue an *assign replenishment* message, which will be handled by the stock planner's *assign replenishment* behaviour. This behaviour will restore the stock level between the minimum and maximum level.

Execution Behaviour. Device managers also have standardised behaviours. An example is the *execution behaviour*, which is triggered by an event from the plant signalling the completion of a task. The execution behaviour will then select the next task to be executed and forwards it to the plant. However, the execution behaviour also takes care of the unavailability of its peers. Before selecting a task to be executed, the execution behaviour first sends its peer device managers a *task inquiry* message. If they are off-line, they will send their lists of unexecuted tasks in a *task reply* message. These task lists will be combined into one list, from which the execution behaviour will select the next task for execution.

3 Implementation

The architecture described in Section 2 has been implemented in Java using Jade [1]. The class diagram for the implementation is shown in Figure 4. A Jade application consists of a number of agents, each having a collection of behaviours.

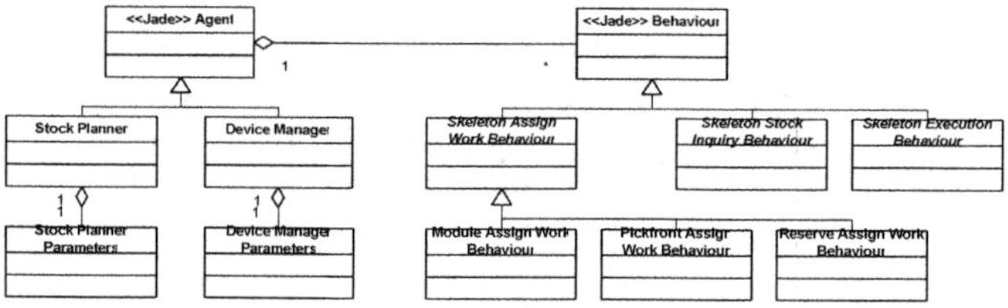

Fig. 4 Warehouse architecture class diagram

Agents. We distinguish two types of agents: stock planners and device managers. These correspond to the components described in Section 2. Both agent types are parameterised specialisations of Jade's Agent class. The parameters of the stock planners include the parent and child stock planners, the connected device managers, the initial stock level, and a collection of behaviours. The main parameters of the device managers are its behaviours and its peer device managers.

Behaviours. The agents' behaviours are the main ingredients for obtaining the desired system behaviour. As described in Section 2, the behaviours have been standardised. This means the interaction between the agents has been fixed, both with respect to the interacting agents and the interface object types. However, the detailed content of the exchanged messages has not been specified. In fact, the behaviours described in Section 2 are skeleton behaviours that have to be made application specific. The skeleton behaviours have been implemented as abstract specialisations of Jade's Behaviour class. The skeleton behaviours declare a small number of placeholder functions, without specifying them. Each agent has concrete behaviours, which are specialisations of the skeleton behaviours. These concrete behaviours fill in the unspecified placeholder functions with application-specific business rules. This is illustrated in Figure 4 for the assign work behaviour; it shows the abstract skeleton behaviour and its specialisations for the module, the pick fronts, and the reserves of the warehouse system shown in Figure 1.

4 Experimental Validation

The architecture described in the previous sections is being validated using a series of experiments. The first experiment of the architecture involved the system in Figure 1. It focused on the delivery behaviour; replenishment was ignored by having sufficient stock in the pick fronts. As replenishment is not required, the reserves and tray miniloads have been omitted. The corresponding warehouse control system has seven stock planner agents, a module and six pick fronts, and six device manager agents, one for each case picker. Each agent has a number of behaviours. The module has four behaviours, including a specialisation of the assign work behaviour described in Section 2. Another important behaviour of the module is the *forward work* behaviour, which forwards the module's work to the pick fronts. Besides an assign work behaviour and a forward work behaviour the pick fronts also have behaviours handling *stock inquiry* messages and *supply cost inquiry* messages. The case pickers have two main behaviours, one handling task assignments and a specialisation of the execution behaviour described in Section 2.

In the experiment, work corresponding to complete pallets gets assigned to the module, which breaks down the work into individual cases and forwards these to the pick fronts. When forwarding the cases, the module has to select one of its child pick fronts. The corresponding business rule is declared in the skeleton forward work behaviour and specified in the module's forward work behaviour. This business rule selects one of the pick fronts based on the expected completion times and characteristics of the stock containers from which goods are delivered. The pick fronts forward the work further to the associated case pickers, which are responsible for the scheduling and execution of the work.

The experiment shows that simple business rules suffice to forward work through a hierarchy of agents, such that device managers receive an equal share of the work. It also shows that peer device managers can take over each other's work. The experiment includes a case picker that is off-line for a period of time. The execution behaviour of its peer case picker ensures that all tasks get executed in a timely manner anyhow. Moreover, the experiment shows that once all case pickers are available again, they receive an equal share of the work again.

The experiment shows that we can obtain the desired system behaviour by configuring a warehouse control system with appropriate business rules. It also shows that we can greatly reduce the system design effort. The skeleton behaviours have only a few placeholder functions; at most four in the experiment. The amount of code implementing the corresponding business rules is small. Less than 15 percent of the code is application specific; more than 85 percent is generic and can be reused in other warehouse control systems by (automated) system configuration.

5 Conclusion and Outlook

In this paper, we have described a new generic warehouse control architecture. Warehouse control systems consisting of stock planner and device manager agents are built by specifying the parameters of the architecture. These include parameters describing the hierarchy of agents and agents' behaviours. These behaviours are specialisations of abstract skeleton behaviours, which are made concrete by specifying application-specific business rules. A first experiment demonstrated that work can be forwarded through a hierarchy of stock planners and device managers and that equipment unavailability can be handled.

The experiment also showed that the generic architecture can greatly reduce the main challenge of designing agent-based warehouse control systems: the specification of agent behaviours. Because of the standardisation, the behaviour design effort is limited to specifying application-specific business rules in generic skeleton behaviours. This comprises less than 15 percent of the code. The rest can be reused by configuring the agents and their interfaces and behaviours.

We are currently implementing the behaviours that were not needed in the first experiment. Subsequently, we will continue the validation of the architecture with experiments involving both delivery and replenishment behaviour. After the validation, we intend to reduce the effort of designing warehouse control systems even further. We will develop a graphical warehouse control editor, with which a warehouse designer will be able to select warehouse agents and their interfaces. Moreover, the designer will be able to select the agents' behaviours and specify the appropriate application-specific business rules.

Acknowledgments. This work has been carried out as part of the FALCON project under the responsibility of the Embedded Systems Institute with Vanderlande Industries as the carrying industrial partner. This project is partially supported by the Netherlands Ministry of Economic Affairs under the Embedded Systems Institute (BSIK03021) program.

References

1. Bellifemine, F., Caire, G., Greenwood, D.: Developing Multi-Agent Systems with JADE. John Wiley & Sons Ltd., Chichester (2007)
2. De Koster, R., Le-Duc, T., Roodbergen, K.J.: Design and control of warehouse order picking: A literature review. European Journal of Operational Research 182, 481–501 (2007)
3. Graves, R.J., Wan, V.K., van der Velden, J., van Wijngaarden, B.: Control of complex integrated automated systems – System retro-fit with agent-based technologies and industrial case experiences. In: Proceedings of the 10th International Material Handling Research Colloquium (2008)
4. Ito, T., Abadi, S.M.M.J.: Agent-based material handling and inventory planning in warehouse. Journal of Intelligent Manufacturing 13, 201–210 (2002)
5. Hiel, M., Aldewereld, H., Dignum, F.: Modeling warehouse logistics using agent organizations. In: Proceedings of the 2nd International Workshop on Collaborative Agents - Research and Development (to appear)
6. Kim, B.I., Graves, R.J., Heragu, S.S., St. Onge, A.: Intelligent agent modeling of an industrial warehousing problem. IIE Transactions 34, 601–612 (2002)
7. Moneva, H., Caarls, J., Verriet, J.: A Holonic Approach to Warehouse Control. In: Proceedings of the 7th International Conference on Practical Applications of Agents and Multi-Agent Systems, pp. 1–10 (2009)
8. Rouwenhorst, B., Reuter, B., Stockrahm, V., van Houtum, G.J., Mantel, R.J., Zijm, W.H.M.: Warehouse design and control: Framework and literature review. European Journal of Operational Research 122, 515–533 (2000)

Manufacturing Material Flow Analysis Based on Agent and Movable Resource Concept

Pawel Pawlewski

Abstract. The paper presents the results of the author's studies which consisted in planning process taking place in an enterprise characterized by manufacturing of complex products (machine building). The idea of the so-called domains of the CIMOSA concept which has been used for the modeling was presented. Planning process based on the multi-agent architecture called VLPRO-GRAPH was extended by the concept of special agent which checks the material flow in the production system. The formal model of this flow was introduced. Finally results of the numerical experiment of analysis and checking process graph are presented.

Keywords: Distributed manufacturing system, Multi-agent systems, Distributed process planning and scheduling, Process modeling.

1 Introduction

The method of the fast analysis of manufacturing material flow was developed by the author of this paper. The method provides a manager with an opportunity to compare the resources available in the production system in a certain time frame with the resources requirement resulting from the production program within the same period. Additionally the method gives recommendations referring to the re-allocation of particular resources. This method is a part of a planning process based on multi-agent architecture. The paper consists of 6 sections, the second of which presents the multi agent approaches to production planning and scheduling. The third one defines movable resources. The fourth section depicts the implementation idea based on CIMOSA domains concept and VLPRO-GRAPH architecture. Results of the numerical experiment are presented in the fifth section. Finally, the sixth one focuses on the plan of future works.

Pawel Pawlewski
Poznan University of Technology
Strzelecka 11, 60-695 Poznań, Poland
e-mail: {pawel.pawlewski}@put.poznan.pl

J.M. Corchado et al. (Eds.): Trends in PAAMS, AISC 90, pp. 67–74.
springerlink.com

2 Multi Agent Approaches to Production Planning and Scheduling

Traditional approaches to production planning and scheduling in MRP based logic do not consider real-time machine workloads and shop floor dynamics. Therefore, there is a need for the integration of manufacturing process planning and control systems for generating more realistic and effective plans. Traditional approaches to production planning and scheduling are based on Centralized Optimization Algorithms, Close Loop Optimization and Distributed Process-Planning (DPP) Approaches. The overview of approaches to agent base production planning and scheduling can be found in [9]. Agent-based approaches provide a distributed intelligent solution by multi-agent negotiation, coordination, and cooperation. The following researches refer to the application of multi-agent systems for production planning purpose [9]:

- bidding based approach – the process routes and schedules of a part are accomplished through the contract net bids;
- a multi-agent architecture based on separation of the rough process-planning task as a centralized shop floor planner from the detailed process planning conducted through agent negotiations.
- based on cascading auction protocol provides a framework for integrating process planning and hierarchical shop floor control.

Agent-based approaches provide a natural way to integrate planning and control, activates and makes it possible to simultaneously the optimize these functions. Some detailed works have inspired our researches. Meta-Morph is an architecture proposed by Maturana et al. [4]. This is an adaptive multi-agent architecture purposed to integrate distributed manufacturing resources for intelligent manufacturing systems. Distributed manufacturing resources are encapsulated as intelligent agents. They are grouped according to the functionalities such as design, marketing, scheduling and material supply. Mediators are special coordinators that are attached to every group. These Mediators connect agents inside the group. An enterprise Mediator is assigned to connect with the Mediators of each group. Preferably, the main agent is principle Mediator who connects the enterprise Mediators together and forms a Mediatorcentric hierarchical architecture, in which scheduling Mediators performed resource scheduling. An agent-based framework to integrate enterprises resources in order to timely plan and re-plan manufacturing processes is proposed by Peng et al.[8]. In this framework different functional agents such as monitoring agent, analysis agent, and data mining agent to deal with the process planning or fault resolving, are included. Holonic concept is used to built an agent-based system for logistics and production planning by Karageorgos et al. [3]. They encapsulate manufacturing units at distributed locations as intelligent agents and used agent negotiation mechanisms to optimize logistics and production planning but this solution does not provide effective resource of alternative strategies.

3 Movable Resources

According to APICS terminology [1], resources are anything that adds value to a good or service in its creation, production, or delivery. Resources are classified as non-renewable and renewable, from the viewpoint of their use, creation and consumption. Non-renewable consumable resources are those which are completely consumed in the production process. Among such resources there are different kinds of materials, forms of energy, products, raw materials, components, auxiliary materials, which are either manufactured in a given company or brought from contractors. As renewable consumable resources we consider machines and machinery, control and instrumentation equipment, human resources, company departments. All these resources are used in the manufacturing process; after use that are released and await another application.

The resources used in the implementation of the process are renewable resources. Renewable resources constitute a system that implements the processes. They are organized as the so-called production structures [2]. That is why, they can be called structural resources. They can be classified as [10]: stationary, exchangeable, free and mobile.

Structural resources are the environment of the process implementation and they create the implementation system of the process. Such a system is a coherent system, which acts as a single entity. Structural resources make a certain supply of resources; among them movable resources can be distinguished. A movable resource is divisible, and either a part of it or the whole resource can be moved between structural units. Examples of movable resources are as follows:

- surface occupied by the implementation system of the process,
- employment measured by the number of employees,
- stocks considered as a full supply of stock in the system (according to the "four wall inventories" concept,
- means of transport assigned to the system.

4 Implementation

Agent-based system is defined in the present paper as a multi-agent system that acts as a support tool and utilizes the databases of main system (ERP system). Multi-agent system is a collection of heterogeneous, encapsulated applications (agents) that participate in the decision making process [5].

The architecture of the tool (VLPRO-GRAPH – Very Long Process Graph) is based on the assumption that the system will support the MPS creation in ERP system and will be plugged in to ERP system database by for example java connector. This architecture was introduced in [7].

Figure 1 illustrates structure and the amount of agents which can be found in each layer. This structure it is an extended structure from [7]. A new agent was introduced, i.e. MR agent (movable resource agent). The main task of this agent is checking the materials domains – D6,D7,D8,D9,D10 for movable resources allocation and limits. The task of this agent is integrated with the planning process which is described at three layers reflecting to [6]: A – the whole process perspective, the so-called whole process planning; B – the entity level where the whole process plan is divided into sub-plans which are executed by each subprocess and being transformed for individual production schedule at the domain level and where local re-planning activities takes place; C - domain sub-layer where production control activities are executed and information about disturbances is gathered and passed to upper levels.

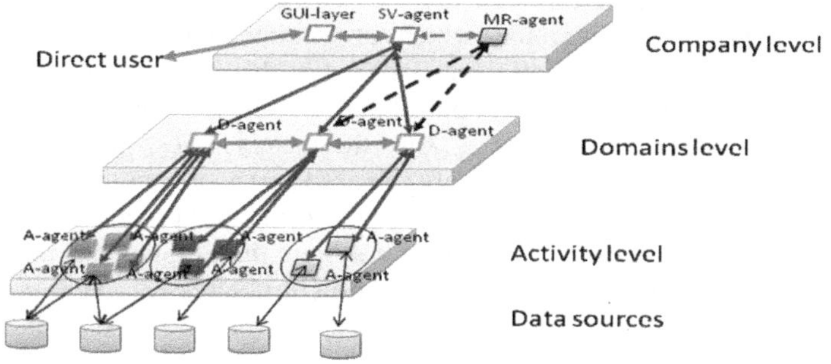

Fig. 1 Extended VLPRO-GRAPH agent model (based on [7])

The graphical user interface agent creates a graphical user interface (GUI) for the interaction of the MAS (Multi-Agent System) to production manager (direct users). The GUI-Agent is able to initialize and send behavior parameters and messages to the Super-Visor Agent (SV-Agent). The SV-Agent is exactly one in the system because the data from all the domain agents (D-Agent) are fused at this agent to generate re-planning schedules for the production. The SV-Agent is responsible for the control of the logic of all agents and creates the plans for the D-Agent. The SV-Agent starts the work of MR-Agent which checks the movable resources allocations and limits. According to the results of MR-agent work, the decision of plans execution is prepared. The planning process is presented in figure 2. D-Agents are initialized by the SV-Agent and they are responsible for the translation of the long process plan into detailed schedules. The agent is allowed to prepare the number of alternative (contingency) local plans as long as they are not conflicting with long process MPS. The local re-planning activities are allowed as long as they don't influence the long process MPS. When re-planning activity affects the long process MPS, it has to be passed to SV-Agent. The A-Agent is responsible for the control of plans execution within sub-process based on the given performance indicators. It reports to D-Agent in the upper layer whether production plans are executed according to given MPS.

The work of MR-agent is based on the operation of adding (shifting) movable resources between structural units called RO. RO (Resource Operation) is a structural unit which contains elements of technological process and resources required to perform this process.

$RO_i = (F', R')$,

Where: F' is a technological process, which consists of a set of the threes describing each technological operation and can be described as:

$F' = (fi, \dots, fml)$ a and each threesome $fi = (tpz, tj, \tau)$ consists of : tpz - setup time, tj - operation unit time, τ - interoperation break; ml - number of operations for 1 unit of assortments

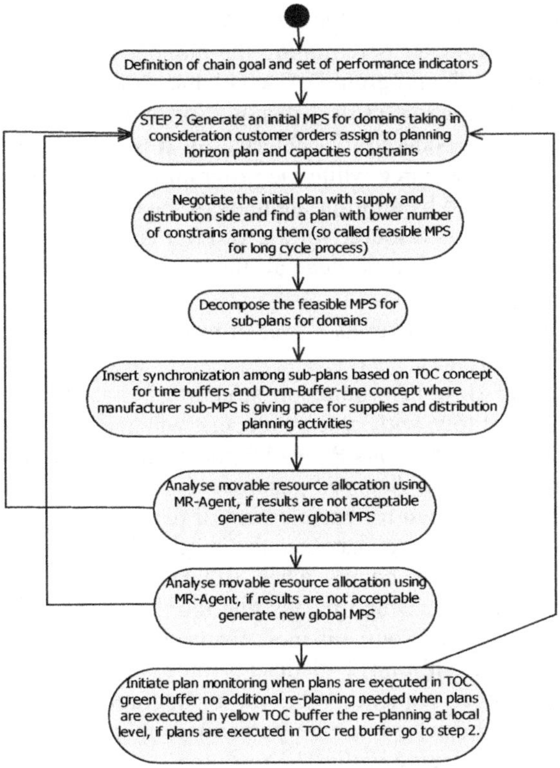

Fig. 2 Extended Planning algorithm for VLPRO-GRAPH agent model

R' is a parameter describing resource requirements for each product offer as an assortment and is described as:

R'= (rm, ... , rn),

and consists of:

R1 – regarding resources requirements of technological operations

$$R1 = [(r1 , \dots , (\dots , ri , \dots)m],$$

$ri = \{zi , \omega i\}$; where zi – type of resource, $i \in <1, n>$,

n – size of resource set, ωi – unit average usage of resource,

m – number of operations

R2 - regarding other resources requirements than R1

$$R2 = [\dots , \{zi , \omega i\} , \dots]$$

$$\bigwedge_{fi} \in ROj \; : \; fi \notin ROk \;\&\; k \neq j$$

If triplets of fi belong to a particular ROj then they cannot belong to any other ROk for each j which is different from k. A unit of assortment in the present paper is defined as a product of different complexity being manufactured within a production system.

Each element of resources requirements R' is connected with a parameter rr – requested resource to implement the production program P in the considered time interval Δt. It is a parameter determining the number of uses of a given resource required to implement the program P. Each element of resources requirements R' is connected with the parameter dr – disposed resource to implement production the program P in the considered time interval Δt. It is a parameter determining the availability of a given resource while the program P is determined though the size of the resource expressed by appropriate units of measure.

The total demand for a resource for a given RO is a sum of individual demands for this resource, requested to produce all the details according to the production program and time per unit.

$ZCi = \Sigma(\omega l * Pl * tl)$ where: ωl - use of resource, Pl - production program, tl – time per unit, $l \in <1, n>$,

The definition of flow in production structure is based on the requested and disposed resources. Flow with reserve occurs when RO has a reserve for any element of resource requirements R', which means that the difference between the required resource and the disposed resource is less than zero. Balanced flow occurs when RO is balanced for any element of resource requirements R' , which means there is no difference between the required resource and the disposed resource. Flow with deficit occurs when RO has a deficit for any element of resource requirements R' , which means that the difference between the required resource and the disposed resource is more than zero.

Operation of adding (shifting) logistics resources symbol :+. For RO, which are in the direct or indirect sequence relation, an operation of adding (shifting) an arbitrary logistics resource rm can be performed only in the situation when the preceding RO have a reserve within the resource.

$$Rm \in ROj = rm \in ROi + rm \in ROj : drm - rrm \in ROi > 0$$

where: s – size of RO files, which are in relation: ROj→ROi or ROj⇒ROi, and have a reserve within the given resource rm, i.e. the condition is met: drm – rrm > 0.

Direct sequence relation symbol:→ ROj is in the direct sequence relation with ROi, which is written as ROj→ROi; if in the production process ROj is a successive element in relation to ROi.

Indirect sequence relation symbol : ⇒ ROk is in the indirect sequence relation with ROi, which is written as : ROk⇒ROi; if in the production process there is only one ROj, which is in the relation: ROk→ROi and ROj→ROi.

5 Numerical Experiment

The research was done based on the data collected for the sake of projects developed by the Institute of Management Technology and ordered by one of machine-building industry enterprises. The investigated object is a typical cooperation company, since all the finished products are units, components and parts which are passed on for further assembly. The company specializes in the production of units and components for agricultural machinery, especially: three-nodes suspension systems for tractors, straw shredders as parts of harvesters, other parts and units (e.g. frames for cultivators). The machinery installed in the enterprise is typical for machine-building companies and it mainly consists of: turning lathes, milling machines, rolling mills, drilling machines, presses, frame saws and guillotine shears, fettling and cleaning machines, semi-automatic welder and automatic machine for oxygen-cutting and assembly, locksmith's and painting stands. These machines form list of groups of uniform stands according to MRP II concept – list of RO. Table 1 presents input data for numerical experiment – technological routines for every produced detail and processes goals.

Based on the data presented in Table 2 the process model was built. To prepare the numerical experiment, special Java application (called ROgraph) was written. The main reason to choose the Java environment was that this environment is the same for Jade (Java Agent Development Framework). Jade is the framework which is used to implement VLPRO-GRAPH [7].

Table 1 Technological routines of production process

Details	A	B	C	D	E	F	G	H	I	J
Process	127	162	175	178	179	219	223	234	235	269
Goal	12000	11000	11000	12000	12000	11000	11000	11000	11000	12000
Operations	4	9	4	4	6	4	6	4	5	5
RO	1	1	10	6	21	2	1	2	2	2
RO	5	19	19	10	2	10	10	10	21	24
RO	2	2	2	14	10	19	25	24	10	10
RO	3	10	3	3	14	3	8	3	24	3
RO		9			12		14		3	21
RO		15			3		3			
RO		19								
RO		3								
RO		24								

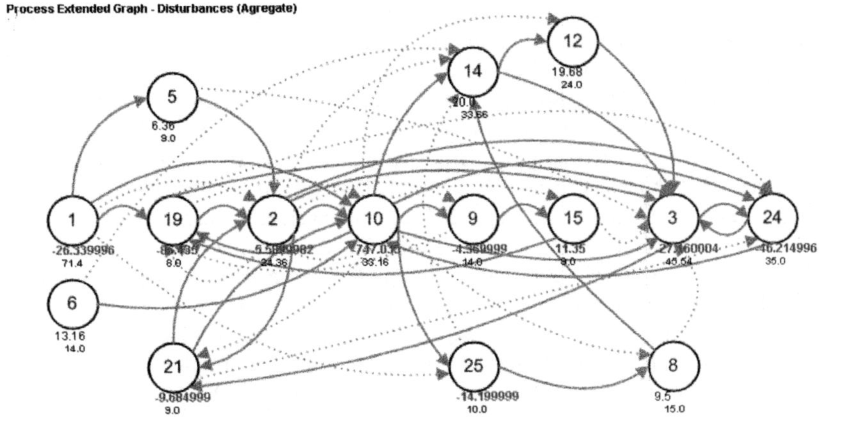

Fig. 3 Extended Planning algorithm for VLPRO-GRAPH agent model

Results of the numerical experiment are presented in Figure 3. ROgraph application generates a full process graph (solid lines present direct sequence relations and dotted lines present indirect relations) and writes numerical results on this graph. Vertex with red numbers indicate that there are flows with deficit and global MPS must be corrected; that means that a new global MPS will be generated.

6 Conclusions

In the present paper, the author presents the extended planning process based on the multi-agent architecture and special agent called MR-agent which checks the potential problems with material flow in the manufacturing system. The results of numerical experiments are presented. The next steps will be performed in two ways: the idea of the first one is to integrate ROgraph application which can present results with VLPRO-GRAPH architecture; the other way is to do research on results interpretation and prepare the scenarios for responding to problematic situations in the manufacturing system.

References

1. APICS: Dictionary, 12th edn. America Production and Inventory Control Society, Inc., Falls Church (2008)
2. Boszko, J.: Struktura organizacyjna przedsiębiorstwa i drogi jej optymalizacji. WNT, Warszawa (1973)
3. Karageorgos, A., Mehandjiev, N., Weichhart, G., Hammerle, A.: Agentbased optimisation of logistics and production planning. Eng. Appl. Artif. Intell. 16, 335–348 (2003)
4. Maturana, F., Shen, W., Norrie, D.H.: MetaMorph: an adaptive agentbased architecture for intelligent manufacturing. Int. J. Prod. Res. 37(10), 2159–2174 (1999)
5. Pechoucek, M., Říha, A., Vokrínek, J., Marík, V., Prazma, V.: ExPlanTech: Applying Multi-agent Systems in Production Planning. Production Planning and Control 3(3), 116–125 (2003)
6. Pawlewski, P., Golińska, P., Fertsch, M., Tujillo, J., Pasek, Z.: Multiagent approach for supply chain integration by distributed production planning, scheduling and control system. In: International Symposium on Distributed Computing and Artificial Intelligence, DCAI 2008, Springer – Advances in Soft Computing, Heidelberg (2008)
7. Pawlewski, P., Kawa, A.: Production process based on CIMOSA modeling approach and software agents. In: Demazeau, Y., Dignum, F., Corchado, J.M., Bajo, J., Corchuelo, R., Corchado, E., Fernández-Riverola, F., Julián, V.J., Pawlewski, P., Campbell, A., et al. (eds.) Trends in PAAMS. AISC, vol. 71, pp. 233–240. Springer, Heidelberg (2010)
8. Peng, Y., Finin, T., Labrou, Y., Cost, S., Chu, B., Long, J., Tolone, W.J., Boughannam, A.: An agent-based approach for manufacturing integration—the CIIMPLEX experience. Appl. Artif. Intell. 13(1-2), 39–64 (1999)
9. Shen, W., Wang, L., Hao, Q.: Agent-Based Distributed Manufacturing Process Planning and Scheduling: A state-of-art survey. IEEE Transactions on Systems, Man and Cybernetics-Part C: Applications and Reviews 36(4) (2006)
10. Zaborowski, M.: Sterowanie nadążne zasobami przedsiębiorstwa. Wydawnictwo Pracowni Komputerowej Jacka Skalmierskiego, Gliwice (2008)

How to Successfully Combine Case Based Reasoning and Multi-Agent Systems for Supply Chain Improvement

Paul-Eric Dossou, Philip Mitchell, and Pawel Pawlewski

Abstract. Case Based Reasoning (CBR) remains widely used for the definition of the needs for the design and development of expert systems. The originality of this article is that it shows how the reasoning is combined with the theory of multi-agent systems. The stages leading to the achievement of the scenarios will be presented together with initial development difficulties.

Keywords: Multi-agent systems, Case-Based Reasoning, Expert system, enterprise modelling, performance, reference models, rules, Knowledge.

1 Introduction

The present international economic climate imposes on enterprises the need for great organization in order to be competitive and improve their performance. For this improvement, enterprise modeling appears as one of the best ways. GRAI Methodology [4] is designed and defined for managing this modeling.

GRAI approach is composed of five phases: an initialization phase for acquiring the context of the enterprise studied, the modeling phase for elaborating enterprise models (functional, physical, decisional, process, informational) according to different points of view, the diagnosis phase for detecting inconsistencies in the models, design phase for improving the models, and the final one being the implementation of the new system.

This paper proposes to develop a tool for supporting this methodology. Different concepts have been presented relating to modules of the global tool. The general structure of the tool is based on Case Based Reasoning (CBR). Indeed, a

Paul-Eric Dossou · Philip Mitchell
ICAM, Site de Vendée, 28 Boulevard d'Angleterre, 85000 La Roche-Sur-Yon, France
e-mail: {paul-eric.dossou,philip.mitchell}@icam.fr

Pawel Pawlewski
Poznan University of Technology, ul. Strzelecka 11, 60-965 Poznań
e-mail: pawel.pawlewski@put.poznan.pl

J.M. Corchado et al. (Eds.): Trends in PAAMS, AISC 90, pp. 75–82.
springerlink.com

Problem Solving Method has been developed in this context for managing the different possibilities of modeling and improvement.

After presenting the concepts of the modules being developed and the combining of CBR with multi-agent systems, the impact on the supply chain will be presented. Then the difficulties of development will also be explained.

2 The Use of CBR

Case Based Reasoning allows to solve new problems by using the knowledge acquired from the old cases [1], [2], [9]. This reasoning is used in GRAI Methodology for defining the structure of a hybrid expert system (fig. 1).

The architecture of this system contains three different bases for managing the study of a new case. A model base is used for managing elaborated reference models. A rule base is used for analyzing the models of the system in question. And a case base is defined for capitalizing different studies for future use.

This system also contains different modules. For each new case, an old case with a similar context will be extracted from the case base, adapted in order to improve the new system (adaptation module and case builder), according to CBR rules.

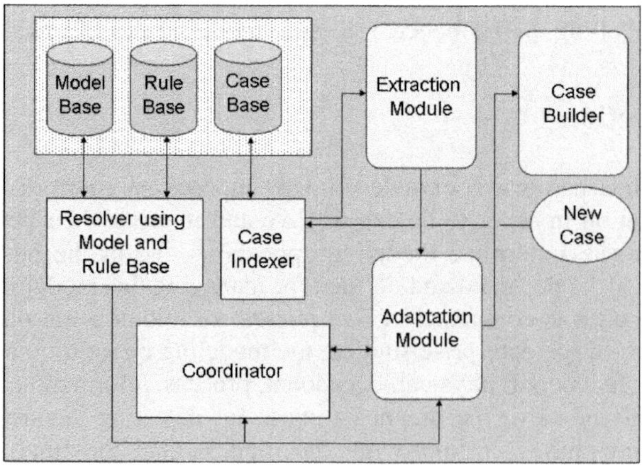

Fig. 1 Architecture of a Hybrid system

This architecture implies the development of a problem solving method associated to the different phases of the methodology [12].

This problem solving method (fig. 2) combines with CBR different types of reasoning (decomposition, direct correspondence, transforming reasoning). The next step after the definition of the tool concepts is the development. The combining of CBR approach with Multi-Agents system theories is proposed. The objective is to obtain an efficient tool for improving enterprise performance and particularly the supply chain performance.

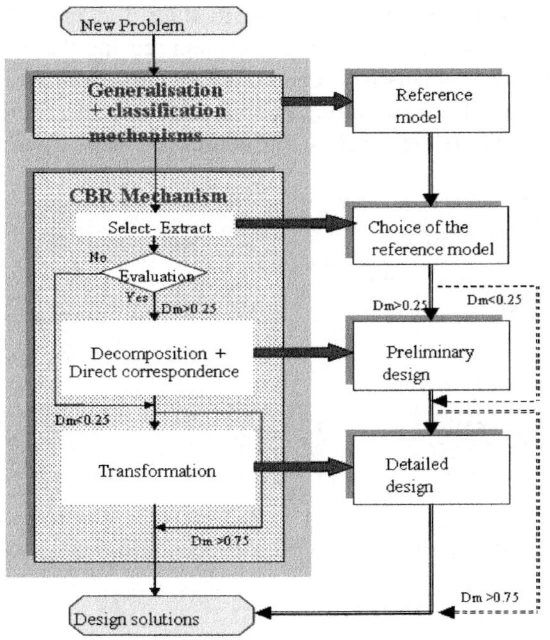

Fig. 2 Problem solving method

3 The New Tool GRAIMOD

In figure 3 the integrated general architecture of the new tool GRAIMOD is presented. We can notice that this tool contains four modules working around a kernel (GRAIKERN). It corresponds exactly to the actual level of design. The techniques used for developing the tool have to integrate a progressive addition of other modules in order to be compatible with the GRAI methodological tree.

GRAIKERN is a graphic editor used for representing the different models associated to GRAI methodology. It is an interface between the different modules.

GRAIXPERT is a hybrid expert system for managing the analysis of the existing system and proposing a new system. It is composed of two sub-modules in interaction with GRAIKERN: the Knowledge Capitalization (KCM) and the Knowledge Based System (XPERTKBM) [3], [4], [5].

The Knowledge based system contains a rule base used for analyzing the different models in order to detect inconsistencies and propose corrections. A dictionary is used to translate user's expressions into standard expressions provided by the GRAI methodology.

The knowledge capitalization process needs some aptitudes to manage different know-how and points of view. It must integrate this knowledge in an accessible, usable and maintainable form. It offers an expertise model based on the knowledge of the experts but also on the previously realised studies. The capitalization module is composed of an acquisition module for integrating other expert knowledge, a case base for capitalizing cases and reusing them during a new modeling, and a reference models base containing models according to different types of enterprise domain. For elaborating the reference models, a production typology is done. This typology is improved by the addition of new criteria.

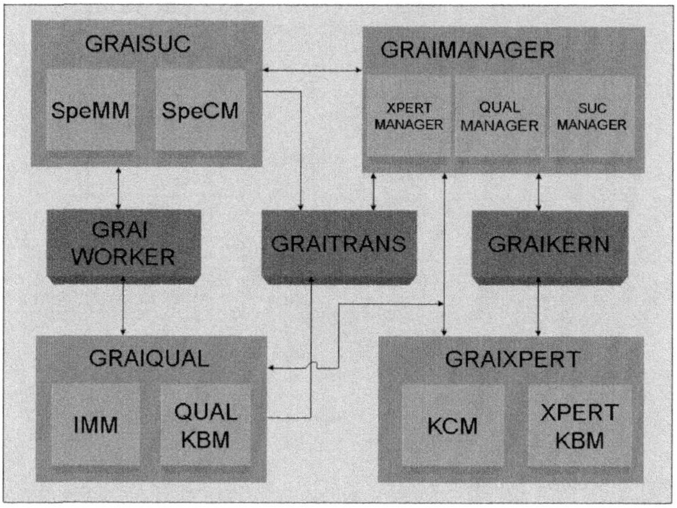

Fig. 3 Architecture of GRAIMOD

GRAIMANAGER is a management module used for organising the different interactions between the modules of GRAIMOD. It controls and manages the system's interactions with the users. It presents the users with appropriate questions and choices together with the necessary information about the characteristic of the enterprise studied. It also manages the rules classified according to a typology of production systems. Its main tasks are the modification, suppression or selection of the applicable rules in a given context. It is also used for the loading and the saving of rule files. Finally, it controls the design process, different actions of the sub-modules and their interactions.

The GRAIWORKER is the work base elaborated for managing, modifying and capitalising knowledge about the studied case. GRAITRANS is a Transfer Interface used for putting the new case in GRAIXPERT in order to improve its Cases Base. The reference model elaborated for each enterprise domain will be improved by the acquisition of this new model in GRAIXPERT [6], [7], [8].

GRAISUC is a module used for managing the choice of an ERP or SCM tool for an enterprise. It is composed of two sub-modules SpeMM and SpeCM The Specification Management Module (SpeMM) is used for choosing the appropriate ERP or SCM Tool of an enterprise. The specifications obtained are capitalised in the Specification Capitalisation Module (SpeCM).

GRAIQUAL is a module used for managing quality approach implementation or quality improvement in an enterprise. It contains two sub-modules IMM and QUALKBM. The Improvements Management Module (IMM) is used for managing the different quality action plans of the enterprise. It contains different quality tools. The Quality Knowledge Base Module (QUALKBM) is being elaborated for containing the rules related to quality certifications in order to use them for improving or elaborating quality in an enterprise.

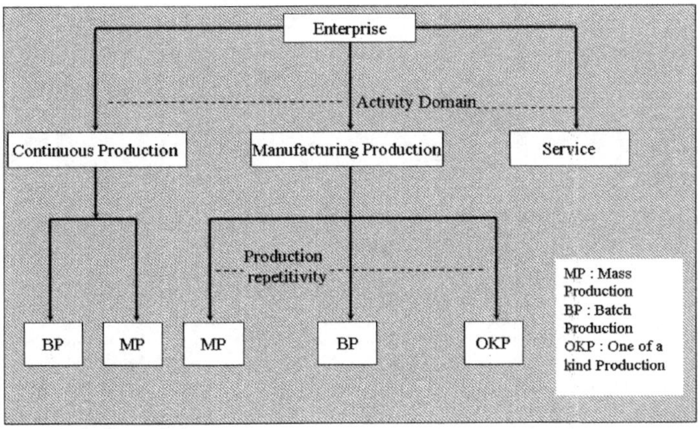

Fig. 4 Typology of enterprises

These different modules are integrated in GRAIMOD. Special attention is paid to the link with multi-agent systems in order to develop this new tool.

4 Combining CBR and Multi-Agent Systems

The architecture of the new tool is based on Case based reasoning. The use of multi-agent systems will allow to facilitate the development of GRAMOD.

Then CBR needs to be related to Multi-agents systems in order to satisfy user requirements. The reactive agents are not appropriate to our problem because they react only for the environment changes. The BDI architecture could be used but we chose the training agents because of their particularity to learn [10], [11]. This system is used for defining how to improve the enterprise performance and for capitalizing the process followed. According to CBR concepts the treated new case could be capitalized, but the parameters would also improve the reference model of the enterprise domain. Java is chosen for developing GRAIMOD. The Jade platform is being used in relation with FIPA-ACL for developing the different modules of GRAIMOD (fig 5.). Some changes could be integrated according to the opinion of Jade specialists.

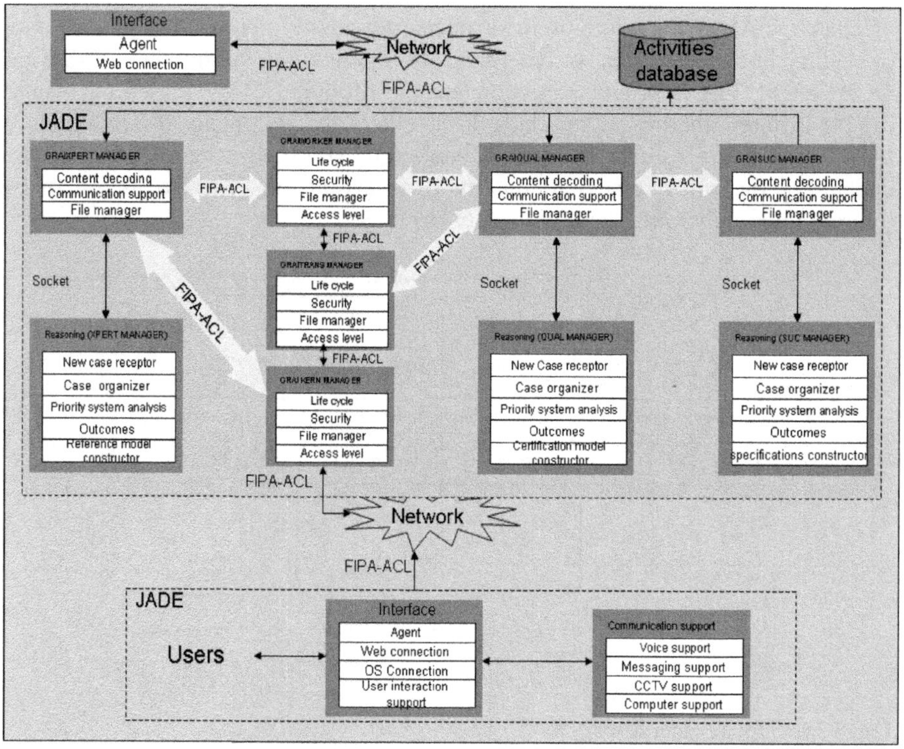

Fig. 5 Architecture of GRAIMOD in a Java environment

5 Global Impact on Supply Chain Performance

The global objective of this research is to be more efficient in the improvement of enterprises. The supply chain of each enterprise could be reorganized by using the concepts elaborated. The reorganization takes into account both the production typology and the supply chain in the modeling management.

For each enterprise, the supply chain is decomposed into different parts (sourcing, procurement, purchasing, production, distribution, sales, transport and logistics management). For each part GRAIQUAL is used and a quality approach is defined in order to improve this part. Indeed, the optimizing of each part is coherent with the other parts.

The importance of the CBR reasoning is evident because of the use of knowledge capitalization realized for each enterprise domain, for each part of the enterprise supply chain. But the necessity of multi-agent systems is to justify this choice. For instance, with the implementation of SQA in an enterprise, not only will the knowledge relating to this enterprise domain be used, but also the system will evolve during this implementation. The system both provides the new case with data and takes into account the particularity of this new case. The multi-agent system defined is well-suited to this kind of work: use and capitalization of knowledge.

Multi-agent systems architecture also facilitates the communication between each different module of GRAIMOD by defining connections from each module to the others. The improvement of quality in the whole enterprise has a positive impact on cost and on delivery date. The integration of environmental demands and the transformation of these constraints into marketing advantage could also be deduced. Then Carbon footprint could be considered as the fourth performance criteria in order to be efficient.

6 Difficulties of Development Phase

At this stage of the project, some scenarios are being defined for facilitating the development of the new tool in coherence with the different factors presented above.

The first difficulty is to translate these concepts into terms appropriate for IT developers. The choice of the use of Java technology and Jade Platform integrates some constraints. The lack of expertise in the use of Java implies the use of specialist skills. The financial aspect of the employment of a specialist is currently being studied. The first idea was to develop different tools with compatibility between each other (GRAISUC, GRAIXPERT …). The meeting with the specialist has changed this view and now the objective is to develop only one tool and different sub-modules (in interaction) of this tool. It means the change of the first scenario of development and the elaboration of another one. It seems that multi-agent systems have to be elaborated separately and then integrated and manage by one tool. The knowledge of experts will definitively define our position.

The last difficulty is related to the type of open tool. By using Java, the objective is to obtain an open tool. But the rights associated to this tool have to be defined. It is difficult to obtain collaboration with a software development company if there is no money for them after the development. The second possibility would be to define and obtain a special research budget for this development.

7 Conclusion

Enterprise modelling is one of the main tools available for enterprises to help them successfully come through the present crisis. The particularity of Icam Vendee (Icam Group), where students spend 50% of their Engineering degree course in industrial apprenticeship, is the great opportunity to obtain applications for concepts developed on Case Based Reasoning and on Multi Agent Systems Theories.

The development of GRAIMOD associated to GRAI Methodology for managing the different phases of this method and improving enterprises will allow to upgrade the expertise given to SMEs for helping them during the crisis and preparing them for the future.

GRAIMOD is being developed in a Java environment, particularly with Jade platform and FIPA-ACL Language.

References

1. Aamodt, A.: Case-Based Reasoning: foundational issues, methodological variations and system approaches. Artificial Intelligence Communications 7(1), 39–59 (1994)
2. Burke, E.K., et al.: Structured cases in case-based reasoning – reusing and adapting cases for time-tabling problems. Journal of KBS 13(2-3), 159–165 (2000)
3. Brown, D.C., Chandrasekaran, B.: Expert system for a class of mechanical design activities. In: Knowledge Engineering in CAD. Elsevier, Amsterdam (1985)
4. Chen, D., Doumeingts, G., Vernadat, F.B.: Architectures for enterprise integration and interoperability. Past, Present and Future. Computers in Industry 59, 647–659 (2008)
5. Dossou, P.E., Mitchell, P.: Using case based reasoning in GRAIXPERT. In: FAIM 2006, Limerick, Ireland (2006)
6. Dossou, P.E., Mitchell, P.: Implication of Reasoning in GRAIXPERT for modeling Enterprises. In: DCAI 2009, Salamanca, Spain (2009)
7. Dossou, P.E., Mitchell, P.: How Quality Management could improve the Supply Chain performance of SMES. In: FAIM 2009, Middlesbrough, United Kingdom (2009)
8. Dossou, P.E., Pawlewski, P.: Using Multi-agent system for improving and implementing a new enterprise modeling tool. In: PAAMS 2010, Salamanca, Spain (2010)
9. Russell, S.J., Norvig, P.: Artificial Intelligence. A Modern Approach. Prentice-Hall, Englewood Cliffs (1995)
10. Sen, S., Weiss, G.: Learning in Multiagent Systems. In: Weiss, G. (ed.) MultiagentSystems: A Modern Approach to Distributed Artificial Intelligence, ch. 6, pp. 259–298. The MIT Press, Cambridge (1999)
11. Wooldridge, M.: Intelligent Agents. In: Weiss, G. (ed.) Multiagent Systems:A Modern Approach to Distributed Artificial Intelligence, ch. 1, pp. 27–77. The MIT Press, Cambridge (1999)
12. Xia, Q., et al.: Knowledge architecture and system design for intelligent operation support systems. The Journal Expert Systems with Applications 17(2), P115–127 (1999)

Market Strategy Choices Made by Company Using Reinforcement Learning

Dawid Chodura, Paweł Dominik, and Jarosław Koźlak

Abstract. This work analyses the results of applying reinforcement learning techniques for chosing agent strategies that model the behaviour of companies within a market. Different kinds of patterns describing states of agent company and strategies describing its activities in given states were analysed. States may depend on company resources (capital, stock level) and activities of customers, suppliers and competing companies. Strategies may be more aggressive or more conservative where the level of profit margin and the range of stock level increases are concerned.

Keywords: company strategy, reinforcement learning, supply-chains.

1 Introduction

Management of production companies need to make decisions concerning different aspects of activities such as production planning, order planning, demand and price prediction, negotiation strategies or stock management. The types of decisions to be chosen for executing and the choice of the appropriate conditions are factors influencing the market success of the company. The complexity of the problem for analysis which require a cooperation of decision mechanisms concerning different domains of the company activities, is an interesting challenge for techniques based on artificial intelligence, and especially multi–agent systems. For these reasons, there is a considerable interest from researchers in these problems, the popularity of "trading agent competition" contests [6] is an example that confirms this.

The goal of the presented work is to develop an environment for modelling and optimising different company strategies [3]. Machine learning techniques for choosing company strategies in identified situation patterns, is presented here. For the

Dawid Chodura · Paweł Dominik · Jarosław Koźlak
Department of Computer Science, AGH University of Science and Technology,
Al. Mickiewicza 30, 30-059 Kraków, Poland
e-mail: pwdominik, dawid.chodura@gmail.com, kozlak@agh.edu.pl

J.M. Corchado et al. (Eds.): Trends in PAAMS, AISC 90, pp. 83–90.
springerlink.com

learning of suitable strategies, the Q-learning algorithm [7] was used, making it possible for an agent to learn an evaluation function of the execution of an action in a given state.

The organisation of the paper is as follows. Section 2 concerns an overview of the domain research. In section 3 the model of the system is described. Some realisation details are given in section 4. Section 5 consists of the description of simulation results and section 6 concludes.

2 State of the Art

Three domains of application of the multi–agent approach for supply–chains are distinguished: modelling, designing and management [2]. The most interesting here is the use of machine learning techniques for improving results of the company. For example, in [5] the reinforcement learning algorithm is used for accepting offers considering their characteristic. The overview of the results obtained using different learning and prediction techniques by company agents during the TAC SCM is given in [4].

In this work, agents use Q–learning algorithm [7] for iterative evaluation of the execution of strategies in given situations. The main scheme of the algorithm is in algorithm 1.

Input: Sets of states (S) and actions (A), discount parameter (γ), parameter (μ)
Output: Evaluation of execution actions in states Q(s,a)
initialise Q(s,a);
repeat
    select action a;
    execute action and receive reward r;
    test new received state s';
    $Q(s,a) := Q(s,a) + \mu(r + \gamma max_{a'} Q(s',a') - Q(s,a))$;
    $s := s'$;
until *simulation finished* ;

Algorithm 1. Q–learning algorithm

3 Solution Description

The developed environment consists of three kinds of agents: clients (which send orders for final products), negotiators (which model the market, assign orders with companies and determine contract conditions) and companies/producers.

The main element for which the strategies and their choices are modelled is a company represented by agent AP^i (agent producer). The goal of the learning process is to set the indices determining strategies according to which the company should act and decide which actions are the most profitable in a given market state. The market state may be dependent on a level of consumer demand, number of parts delivered by suppliers and contract conditions expected by them, the level of competition of the companies with similar production profiles as well as the internal state of the company arising from its capital and stock levels. The learnt strategies may be stock strategy (for example, to buy parts for final production only after receipt of orders from clients or to keep adequate stock levels), required level of profit margin, willingness to conclude long term contracts and the importance of elements associated with this: prices, delivery times or quality.

Agent producer AP^i, which represents the company i, for this work may be represented by the following n–tuple:

$$AP^i = (CA^i, C^i, I^i, CO^i, PD^i, EA^i, L^i, Dec^i),\ L_i = (\phi^i, A^i, S^i, s^i, Q^i)$$

where: CA^i – capital, C^i – daily upkeep costs and inventory unit costs, I^i – states of inventory stocks, CO^i – accepted for realisation orders of clients, PD^i – predicted demands for goods, EA^i – external actions which result in changes of the inventory state or flow of money and goods with other agents , Dec^i – decision function which selects the external action for the execution L^i – reference to learning module (empty if simple not–learning agent), ϕ^i – function which mapping information owned by agent to the state for learning algorithm, A^i – set of possible actions, S^i – set of possible states together with their description, possible states are Cartesian products of sets which consists of possible values of state factors, s^i – current identified state, Q^i – evaluation function of actions performed in given states, calculated using Q–learning algorithm.

The ϕ and Dec functions are defined as follows:

$$\phi^i : (CA^i, C^i, I^i, CO^i, PD^i) \rightarrow S^i$$
$$Dec : (CA^i, C^i, I^i, CO^i, PD^i) \rightarrow EA^i \text{ if } AP^i \text{ not learning agent}$$
$$Dec : (s^i, Q^i) \rightarrow EA^i \text{ if } AP^i \text{ learning agent}$$

Consumer. The consumer is represented by parameters necessary for confirmation requirements concerning the contracts such as demand level, acceptable terms of contract, utility function.

Negotiator. The task of the negotiator is to assign consumer orders and producer offers and determine conditions of agreed contracts. It is assumed that in each stage, agents have a given number of bids to start contracts.

The buying order (BO^j) is described as follows:

$$BO^j = (PN^j, n_j, mq^j, mp^j, mdt^j, \alpha_p^j, \alpha_q^j, \alpha_{dt}^j)$$

where: PN^j – name of the product, n^j – number of units, mq^j – minimum quality of product, mp^j – maximum price of the unit, mdt^j – maximum accepted waiting time for delivery, consumer preferences – three weights α_p^j, α_q^j , α_{dt}^j, which inform how important price, quality and delivery time are for consumer respectively.

On the basis of an order the negotiator formulates an invitation to negotiate – NI^j for producers manufacturing a given product. This is described as follows: (PN^j, n^j, mq^j, mdt^j) where: PN^j – name of product, n^j – number of units, mq^j – minimum quality, mdt^j – maximum delivery time.

The response to the negotiation invitation sent by the producer to negotiator the producer PR^j is represented by the triple (p^j, q^j, dt^j) where: p^j – price of the unit, q^j – quality, dt^j – delivery time.

After receiving all offers form the producers, the agent negotiator makes a choice of the winner on the basis of preferences sent by a customer. The producer with the highest score is the winner.

4 Realisation

System description. System provides three types of services: *Information* (supply

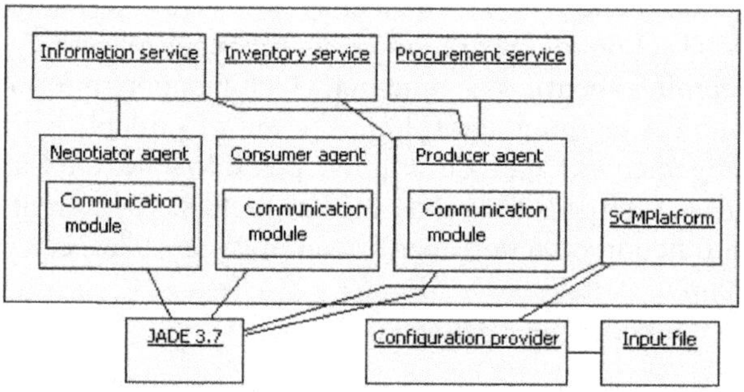

Fig. 1 System concept

basic information about agent configuration: production capability, initial capital, inventory unit cost), *Inventory* (supply information about producer warehouse status), *Procurement* (predict future parts usage based on orders history). On the realisation level the system consists of one additional agent – *Simulator*. Each agent uses *Communication module* to send and receive messages. Using *Communication module* makes system independent from underlying messaging layer – JADE [1].

The simulation process is divided into days and each day into phases: order phase, delivery phase, gathering phase. In *order phase* Consumer places orders. Producers may also send their orders and offer responses to other producer orders. During the *delivery phase* producers send products to consumers if they have manufactured all products of a specific order. The *gathering phase* consists of producers collecting all shipments and storing them in their warehouse.

Negotiation process. The negotiation process is a part of the order phase. In the tests it is assumed that in each order phase there are five steps of negotiation processes. The following sub–phases are defined: (1) collecting orders from all producers and Consumer, (2) sending invitations to producers, (3) collecting offers from producers,

(4) choosing the best offer for each order, (5) sending messages about negotiation results to clients and producers.

Learning process. A Q-learning algorithm needs a set of states that describe the agent environment. The market description was decomposed into independent parameters called "state factors". The independence of state factors in real-life is hard to achieve as it is nearly impossible to predict the influences between values of state factors. The motivation for using an independent state factors set is to reduce the size of the Q-learning table. Each state factor has a discrete set of values. The final state is built up from a set of state factors and their values.

5 Results

Below, a selection of performed experimental scenarios is presented. The first one concerns the comparison of the results obtained by agents using prediction techniques with agents using reinforcement learning techniques. The following concerns the comparison of results obtained by learning agents having different definitions of states and strategies for different market states described by the dynamics of the demand values of the customers.

Comparison of the producers using machine learning and using only prediction algorithms. Five producers participated in the experiment. Only four of them *Producer (P1), LearningProducer1 (LP1), LearningProducer2 (LP2), LearningProducer3 (LP3)* were at the second level of the supply chain analysed during the tests, they were competing among themselves to sell product *Pr3* to Consumer. One producer was at the first level of the supply chain producing two parts that were used in manufacture of *Pr3*.

Table 1 Experiments. States of agents.

State factor	Description	Possible values	Agents – exp. 1	Agents – exp. 2
S_{NO}	obtained new order	yes, no	LP1	LP2
S_{SL}	stock level	low, optimal, high	LP1, LP2, LP3	LP1, LP2, LP3
S_{LN}	lost negotiation	yes, no	LP2	LP3
S_{RN}	negotiation results	won, lost, no		LP1
S_{PWN}	perc. of won negot.	<25, $<25,75>$, >75	LP3	

In the tests, *P1* was an agent using simple algorithms without machine learning. Three learning agents (*LP1*, *LP2*, *LP3*) used different states and action configurations: $S^{LP1} = \{S_{NO}, S_{SL}\}$, $A^{LP1} = \{A_{IM}, A_{DM}, A_B, A_N\}$, $S^{LP2} = \{S_{SL}, S_{LN}\}$, $A^{LP2} = \{A_{IM}, A_{DM}, A_B, A_N\}$, $S^{LP3} = \{S_{SL}, S_{PWN}\}$,
$A^{LP3} = \{A_{IM}, A_{DM}, A_B, A_N, A_{IMS}, A_{DMS}\}$. The state factors and strategies of the agents are described in tab. 1 and tab. 2. The states and strategies of agents in these

Table 2 Experiments. Actions of agents.

Action	Descriptions	Agents – exp. 1	Agents – exp. 2
A_{IM}	increase margin	LP1, LP2, LP3	LP2
A_{DM}	decrease margin	LP1, LP2, LP3	LP2
A_B	buy parts	LP1, LP2, LP3	LP1, LP2, LP3
A_N	do nothing	LP1, LP2, LP3	LP1, LP2, LP3
A_{IMS}	increase margin significantly	LP3	LP1, LP3
A_{DMS}	decrease margin significantly	LP3	LP1, LP3

tests are defined in columns *Agents – exp. 1* of these tables. The learning algorithm of *LP1* and *LP3* agents gives positive reward for buying parts in cases of deficiencies in the warehouse and increases the margin when the producer wins negotiations. It gives negative reward for buying parts when there is excess in the warehouse and decreases the margin when the producer wins negotiations. For *LP2*, reward function was equal to producer capital at that moment.

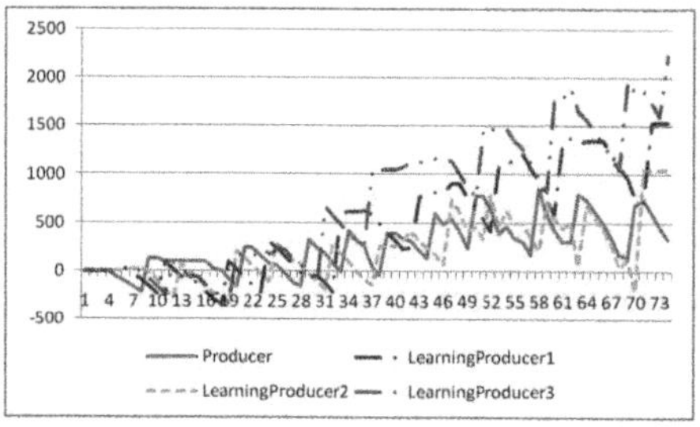

a) balance with parts

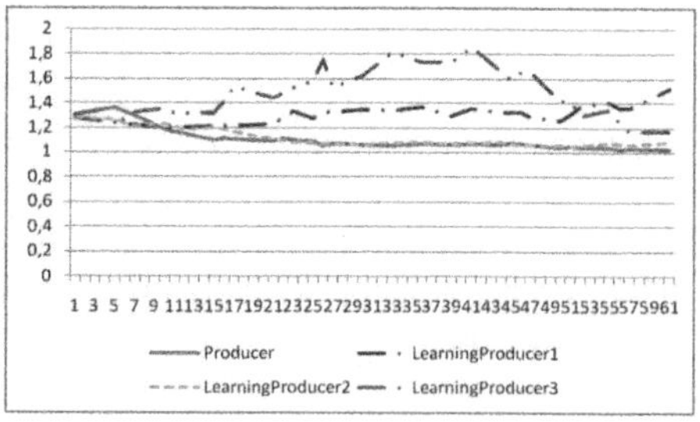

b) margins

Fig. 2 Results of producers using the machine learning and producer using simple algorithms

LP3 won the simulation. It had the most complex action set and state set. It also had the highest margins (fig. 2b)) during the simulation. *LP3* sold less, but higher margins resulted in more profits. *LP1* had very similar behaviour to *LP3*, it also gained good financial results and had high margins.

Differences among producer strategy in different external demand. Experiments of this group had the same producer configuration. The difference between experiments is in different external orders. Producers which were using machine learning had been learning using state and strategy schemes described in columns *Agents – exp. 2* of tabs. 1 and 2. Their sets of states and actions were as follows: $S^{LP1} = \{S_{SL}, S_{RN}\}$, $A^{LP1} = \{A_B, A_N, A_{IMS}, A_{DMS}\}$, $S^{LP2} = \{S_{SL}, S_{NO}\}$, $A^{LP2} = \{A_B, A_N, A_{IM}, A_{DM}\}$, $S^{LP3} = \{S_{SL}, S_{LN}\}$, $A^{LP3} = \{A_B, A_N, A_{IMS}, A_{DMS}\}$.

The producers compared made only one type of product. *LP1* has the most complex sets of actions and states. State set covered all possible situations in selling and warehouse management. Producer could modify the margin more significantly and respond faster to market conditions. *LP2* did not have complex information about negotiation results, it could not distinguish between a situation when it had lost the negotiation and when there were no negotiations. Finally, *LP3* did not have complex information about negotiation results, as it could not distinguish between situations when it had won negotiations and when there were no negotiations. Producer could modify margin more significantly and respond faster to market conditions.

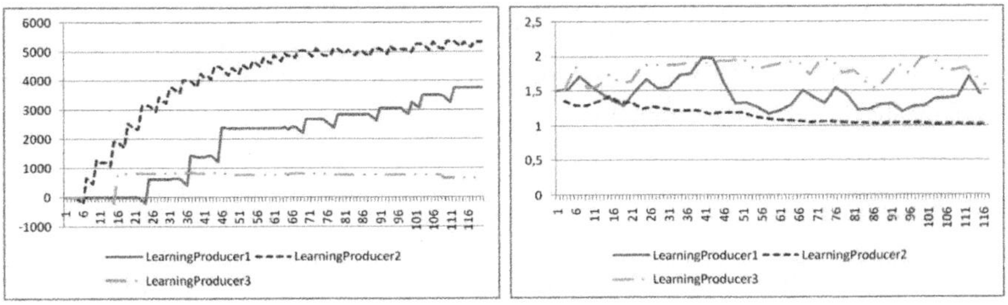

a) balance with parts b) margins

Fig. 3 Results gained on learning data set

Evaluating learning on the learning data set. Consumer orders had a small number of units, so they could be done in a short time and the most important criteria in the negotiation process was the price. *LP2* gained the best financial result (fig. 3a), won most of the negotiations, because it had lower margin than other producers (fig. 3b). *LP1* achieved the second financial result, and *LP3* gained the worst.

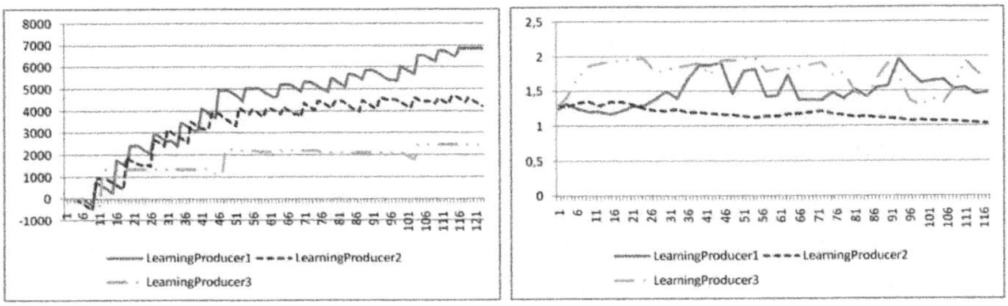

a) balance with parts b) margins

Fig. 4 Results gained on quasi random demand

Evaluating learning on quasi random demand. In this experiment where external orders had been quasi random in terms of product quantity, the best results were gained by *LP1*. It had the most complex action set and state set, and it gave it an advantage over other producers. *LP2* gained the second financial results, it had the lowest margin (fig. 4b)) in most days of simulation, so it could offer the lowest price. *LP3* and *LP1* had higher margins than *LP2*, but they win negotiations because of a shorter delivery time (when *LP2* had busy production lines).

Experiment conclusions. *LP1* was built with the intention that it will win all the tests, as it had complex information about negotiations, but the experiments showed that the initial supposition was wrong. *LP1* learnt that, if it is winning negotiations, then it should increase margins, if it is losing negotiations, then it should decrease margins, if there were no negotiations then it should do nothing, and that strategy seemed to be logical. But when there were many orders with a high quantity in a short amount of time (better to win less negotiations but with higher margins), strategies of *LP2* and *LP3* proved to be better, as they were able to win negotiations offering shorter delivery times than *LP1*.

6 Conclusion

In this work the choice of the company strategies was based on the techniques of reinforcement learning which, when applied, made it possible to improve the obtained results. The selected situation patterns identified gave different results depending on the situation of the analysed market defined by the distribution of demands in time.

Future works will concern the influence of different situations on the market on the optimal definition and identification of states used during the learning process. Another domain of future work is an integration of the analysed solutions with the market environment with more diverse possibilities of the choice of algorithm for company management.

Acknowledgements. The presented work has been financed by a grant from the Polish Ministry of Science and Higher Education number N N516 366236. We would like to thanks J. Marszałek and P. Kolarz, the former students of Dept. of Comp. Sci. AGH-UST for the development of the initial version of the system.

References

1. Java Agent Development Platform (2008), `http://jade.tilab.com`
2. Moyaux, T., Chaib-draa, B., D'Amours, S.: Supply chain management and multiagent systems: An overview. In: Chaib-draa, B., Muller, J. (eds.) Multiagent Based Supply Chain Management. SCI, vol. 28. Springer, Heidelberg (2006)
3. Nawarecki, E., Koźlak, J.: Building multi-agent models applied to supply chain management. Control and Cybernetics 39(1), 149–176 (2010)
4. Pardoe, D., Stone, P.: Adapting in agent-based markets: a study from TAC SCM. In: AAMAS 2007 (2007)
5. Stockheim, T., Schwind, M., Koenig, W.: A reinforcement learning approach for supply chain management. In: 1st European Workshop on Multi-Agent Systems (2003)
6. TAC SCM Game Description (2008), `http://www.sics.se/tac/`
7. Watkins, C.J.C.H., Dayan, P.: Q-learning. Machine Learning 8(3), 279–292 (1992)

Myopic Behaviour in Holonic Multiagent Systems for Distributed Control of FMS

Emmanuel Adam, Gabriel Zambrano, Cyrille Pach,
Thierry Berger, and Damien Trentesaux

Abstract. Multi-Agent Systems (MAS) are particularly adapted to deal with dynamic distributed environments and are typically used to manage business workflows or data flows in manufacturing systems. But, the lack of a global view and reference of the whole system by the low-level and physical parts, to actually guarantee optimal decision-making had been labelled by some researchers as 'Myopia'. Thus, this paper intents to identify some Myopias, involved in a decentralised management of production systems. A HoloMAS (Holonic MultiAgent System), whose elementary parts show signs of myopia, is defined; and we validated our HoloMAS proposal using simulations and through a real implementation on a flexible assembly cell in our university lab.

Keywords: Multi-Agent System, Roles, flexible manufacturing systems, holonic manufacturing system, myopia.

Emmanuel Adam
Lille Nord de France, F-59000 Lille, France

Gabriel Zambrano
UVHC, TEMPO Lab., 'Production, Services, Information' Team, F-59313 Valenciennes, France

Cyrille Pach
Pontificia Universidad Javeriana, Department of Industrial Engineering, Bogotá, Colombi

Thierry Berger
UVHC, LAMIH, F-59313 Valenciennes, France

Damien Trentesaux
CNRS, UMR 8530, F-59313 Valenciennes, France
e-mail: emmanuel.adam, gabriel.zambrano, cyrille.pach,
thierry.berger, damien.trentesaux@univ-valenciennes.fr

J.M. Corchado et al. (Eds.): Trends in PAAMS, AISC 90, pp. 91–98.
springerlink.com

1 Introduction

Manufacturing firms also need constantly be flexible and adapt to newly developed processes and technologies and to rapidly changing environmental protection regulations. To deal with these challenges, one of the new architectures for managing Flexible Manufacturing Systems (FMS) is holonic. Indeed, Holonic Systems are typically one of the new architectures to manage FMS; a consequent number of model, of methods and of implementations have shown their efficiency.

A holonic organisation is a pyramidal organization having specific working mechanisms, described initially in [6]. One of the particularity of holonic concept is the notion of control, which is both distributed on each member of the system, and partly centralized or managed at a higher level. This typical control is particularly interesting in the management of systems evolving in dynamic and demanding environment. A holonic entity call for help to its supervisor only in case of severe disturbances or constraint. Indeed, in each entity, decisional capabilities give them a sense of equality, without a sense of a master-slave relationship among them, and empowering independence.

However, these architectures bring new inconveniences and numerous advantages, well documented in [11]. So a decision is made by independent entities (the principle of autonomy), and a myopic behaviour appears, that is well-known in MultiAgent Systems where agents have to make choices according to volatile and local knowledge (self-knowledge of itself and its surroundings).

The following section present the different notions of Myopia and some approaches used to define robust systems under different degree of Myopia. Next, we present briefly the key concept that we use to build our Holonic multiagent systems (section 3). An application of this concept is then presented to illustrate our approach (section 4).

2 Approaches for Reducing the Impact of Myopia

We can find the notion of myopia in several domain, with different definition. But we can summarise these myopia in he following types:

- Temporal myopia: this first type of myopia deals with decisions in time and is related to the willingness to follow a planning over a long-time horizon.
- Spatial myopia: this second type of myopia induces the fact that an entity/an agent has a limited view of its environment.
- Capability myopia: is related to the ability of an entity to re-configurate its competencies to build new value propositions, and to tackle rapidly changing environments [5].
- Boundary myopia: deals with the fact that an entity may always communicate and work with the same group of entities [5].

Approaches found in literature, to diminish the impact of myopia within FMS can be classed in two groups: the vertical approaches' group, and the horizontal approaches' group.

The *vertical approaches*, include certain elements of hierarchical systems, having some sort of 'superior' entity that will provide a global perspective (in time and space) over the decision-making processes happening on a 'lower level'. Depending of how strong the hierarchy is, the supervisors act permanently, periodically or sporadically. Several layer of control can interact (or even negotiate) between them in order to reach the global objectives (see [4] for instances). Individual objectives are achieved based on constraints released by superior levels and equally, superior entities take into consideration lower entities' goals and systems performance when making decisions. Less restrictive, in ADACOR [7] the concept of a supervisory entity include not only the global production optimisation, but also the reorganisation capacity of the MAS: when the system performs within the boundaries, a heterarchy architecture is deployed; in the case of hard disturbances or instabilities, the architecture flips to a static hierarchy, in which a supervisory entity takes over the control of the situation and solves conflicts, imposing orders to maintain coherence among local and global objectives.

The *horizontal approaches*, horizontal approaches prefer to empower autonomy, embedding 'prediction and supervision' functionalities within entities, improving decision-making by evaluating future states and/or finding optimal choices as a result of negotiation and cooperation, without having any kind of hierarchy status. To avoid local inter-blocking between identical entities, some agents could be enhanced by a role of mediator, able to simulate a future state of the system in order to guide the other agents. Whether these capabilties are embedded into staff holons as shown in [3] or in Autonomous Control Entities (ACE) as in [10] and [9], data must be availabe, synchronization must be assured and simulation times and scalability must be controlled [8], in order to achieve well established dynamics.

A third group of approaches could be found; the hybrid approaches' group where approaches, like in holonic architecture, use both a vertical and horizontal properties of decision-making. Following holonic concepts, we propose a MAS definition particularly adapted to the of multi-levels notions of control, behaviour and myopia.

3 Proposition of a Multi-Level MAS Using Roles

We have analysed holonic rules proposed in [6] and we proposed a first set of formal definitions of a multi-agent system ([2]) having a pyramidal architecture and whose agents have to reach global objective of the system while having a personal goal.

As holonic system is composed of several layers of more or less complex agents (having more or less degree of autonomy, of strategic/reactive behaviour), we tried to define a generic model of agent, that we hope applicable to a large spectrum of MAS application (implying a limited number of deliberative agents).

We present in this article a subset, relative to the notion of autonomy, of these definitions by focusing only on the notion of agent and role.

MAS and agent definition: MAS is simply defined as a set of agents. Each agent is composed of (cf. eq. 1): states; knowledge (social (name of the acquaintances), environmental and personal (which contains the agent goal and agent properties dependent on the application)); messages; a perception function; behaviour rules (functions modifying agent states according to current states, knowledge and received messages in order to reach the collective goal while moving toward its personal goals); a dynamic list of roles that define some behaviors relatively to the application.

$$agent_a = \begin{pmatrix} Knowledge, HRA, messages_a, \\ perception_a, rules_a, roles_a \end{pmatrix} \tag{1}$$

Roles and rules definition: An agent can: receive a role; leave a role; delegate a role or goal to its acquaintances. In our case, we consider a role as a functional role (not as an organisational role) with: a set of knowledge (environmental and social); and a set of rules that allows it to reach its goals.

So, the definition of a role R is, for us, relatively similar to agent definition; a role gives to an agent: the data (environmental knowledge) necessary to accomplish the role; the acquaintances (name of the roles that are linked to this role); the working rules (*hardRules*) in which the agent has to choose a rule in order to be considered as playing this role; a set of rules ($flexibleRules$) able to complete the role. For a role *ro*, each rule R is composed of a set of tasks to execute.

The agent behaviour consists in acting as long as the collective goal, the roles goals and finally the individual goal are not reached; namely, as long as the states of the agent and of its roles does not correspond to the constraints in the Collective Goal CG and the Individual Goal GI. This order is arbitrary, and can be changed, but it corresponds to the notion of cooperative agents that is, we think, essential in the context of a FMS management.

Applying a rule task can lead: to the modification of the personal, environmental and/or social knowledge; to the sending of a message; to the creation of an agent; to the delegation of roles and goals; and to the change of responsible agent.

A rule is chosen relatively to the intended goal and according to the priority (utility level) of the rule. These priority is either defined a priori, or computed from heuristics, or infered from the last use of the rule (the rule's priority increases according to its capacity to go near the agent's goal, each time the agent chooses that rule to reach its goal).

Multi-levels of control/Myopia: To respect holonic principles, the cardinalities of the set of rules and of the set of roles of an agent decrease according to its level (the level 0 being the base of the organisation). The more an agent is at the bottom of the organisation (the more it is close to the physical devices), the less its number of strategies, number of different behaviours is important, and the more its actions are productive (opposed to deliberative). This allows the agents at the level 0 to react rapidly and precisely. On the other hand, the more an agent is at the top of the system, the wider is its choice among several roles and rules; and the more its actions consist in guiding activity of agents below it.

So the myopia, whether it is a boundary (limited view of its environment) or a capability (limited choice of action) myopia, is more important at the bottom of the system than at its upper levels. This particularity is absolutely essential for the efficiency for a MAS that manage a FMS. Indeed, agents linked to the physical parts must react rapidly to find a solution to an eventual failure; being not limited by myopia, these agents will lost crucial time to search and adapted solution (by analysing a too large environmental, social knowledge, by checking the pertinence of a larger number of alternatives).

But Myopia of low level agents could lead to a inter-blocking; if such a situation is detected by the agent, they can inform their supervisor that then search for a solution at its level, by having a larger view of the environment and by interacting with other supervisors at its level.

However, this recourse to the higher level must be relayed as long as possible, because it has a cost (in term of time, messages exchanged), and the agents must find a way to correct their myopia by themselves as long as possible.

We show in the next section a way to improve the potential field mechanism of control.

4 A Case Study for Solving Spatial Myopia

In this section, we propose to illustrate (and reduce) the myopia phenomenon, when controlling the AIP PRIMECA FMS (Flexible assembly cell located at the university of Valenciennes) under a potential fields' implementation described in [12].

In our scenarios, products P_i are placed on shuttles, enhanced by a eeePCs, that have to reach resources. The assembly cell contains several divergent nodes, thus allowing agents in the eeePC to choose between two routes. Potential Fields are devoted to the dynamic routing and allocation processes; here, resources emit a potential field related to their attractiveness (the bigger the field, the more interesting becomes the resource). Fields are reduced if a product enters the resource or its waiting queue. Then, fields are propagated throughout the cell and weakened by the distance. Products are able to sense the fields when they reach a Decisional Node (DN, in fact, the divergent nodes where a product can interact with other entities, makes a decision and communicates it). Each field is emitted for one service provided by the resource, so products have to focus on the fields related to the service they need.

A simulation has been implemented with Jade, based on a previous simulation that we have develop in the context of open-control concept [1].

We give here only two scenarios in order to illustrate a proposition to improve a myopic system by decreasing the temporal myopia. Two products P_1 and P_2 are respectively on the shuttles A and B and have to be processed by the workstation W_1 and W_2 (here, the two workstation propose the same processing). These workstations, considered as resources, emit a potential field.

On the Figure 1a (on the left) the shuttles have interacted with the divergent node on the left of the cell, and have decided to go to the workstation W_1, attracted by the weight of the potential field emitted by it. However, the best strategy for the shuttle B would been to choose to go to the second workstation W_2. Indeed, A and B are too close, when A arrives on W_1, the potential field of this workstation is decreased, but it is too late for B corrects its choice; B must wait until A completes its processing (cf. Fig. 1b). However, if a third product arrives in the top of the cell when A is on W_1, this product will choose to go to W_2, whose the potential field is more important. In this solution, we have implemented two main roles: a potential-field-sensible role and a workstation-responsible role for each agent linked to a workstation.

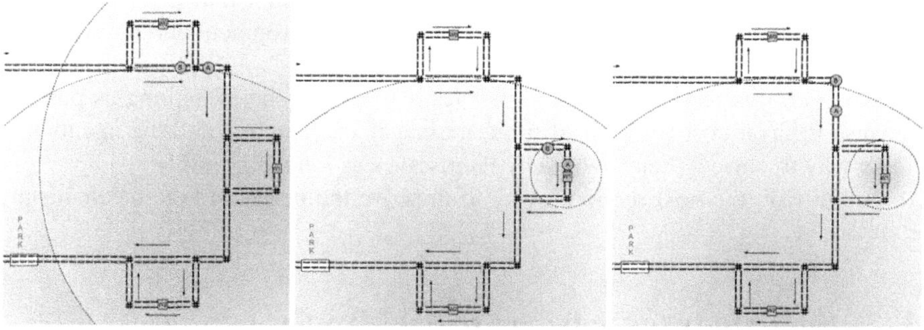

Fig. 1 Influence of potential fields (extract of the case study)

To tackle this problem, a first solution with the objective of reducing myopia without sacrificing autonomy and heterarchy has been proposed: by having all local resources available, meaning that a product/a shuttle will know its environment by reading a table, diminishing spatial myopia; by allowing products to know what will be the state of the system in the next window frame, making them to explore other possibilities and guide themselves not only based on potential fields' levels, but also on the knowledge of other products' choices.

With this solution, when the shuttle A decide to go to W_1, it informs an information layer (simulated by a responsible agent). On Figure 1c, when the shuttle B arrives to the decisional node on the right, it interacts with the information layer (its responsible agent) and is informed about all the available workstations, and about all the 'contracts' (in fact relations) passed between workstations and shuttles/products. B filters the potential field coming from W_1 and decides to go to W_2. When A leaves W_1, it informs the information layer that removes the relation between these two elements from its data space (knowledge for the responsible agent). In this solution, we have now implemented a third role: a cell-responsible role.

Two experiments where deployed at the AIP PRIMECA FMS cell [12] (cf. Fig. 2) of our university, representing the before and after myopia correction.

Gantt diagrams with their results are presented below, in Figure 3. Here it is clear that myopia causes a queue in resource 1 (W_1), enlarging the time for Product 2 (on B) to accomplish the process (84 seconds in total). On the contrary, in the second case, the information unit helped P_2 to decide going over Resource 2 (W_2) allowing it to finish just right after Product 1 (on A)(67 seconds in total). It is important to emphasize that W_2 is located farther from the decisional node DN than W_1, and closer to the final destination node.

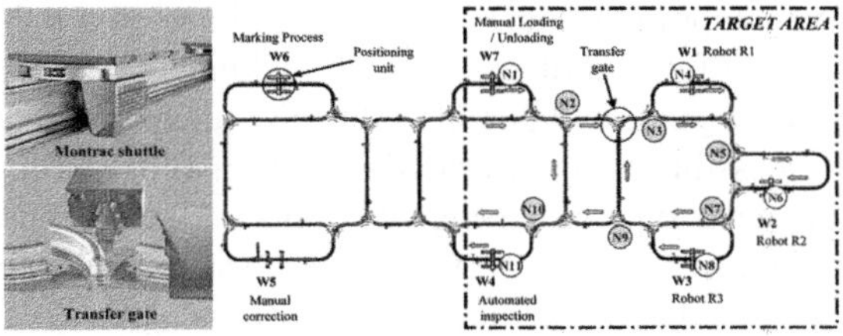

Fig. 2 The flexible assembly cell

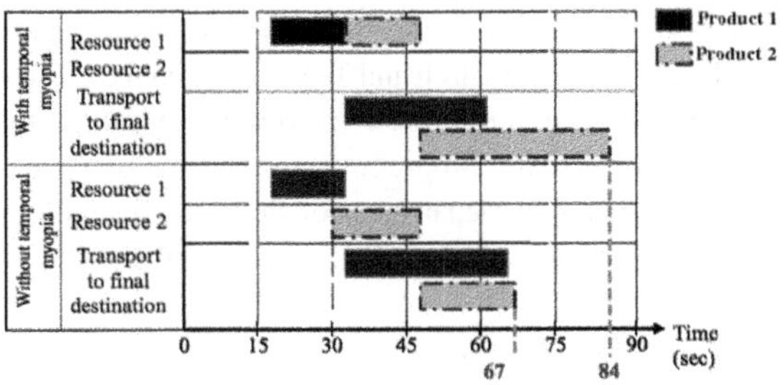

Fig. 3 Gantt Diagram showing the case study results

With this basic approach, we have won approximately 17 seconds over an 84 seconds production time, which represents 20% of time within a production order including just two products. Without modifying the architecture and making huge efforts, we have been able to reduce significantly myopia and increase robustness to uncertainty.

We have shown that some myopic behaviours of autonomous components could be improve by adding a level/layer that allows them to be aware of their local context. The next step of or approach is to study the scalability of the myopia when number of products, resources/workstation drastically increases.

5 Conclusion

FMS control architectures experiencing relationships among autonomous entities have demonstrated their benefits in terms of reactivity to disturbances, self-organization and cooperation to achieve better responses face to internal and external changing environments.

But this type of control architectures have also shown some drawbacks due to the fact that there is no global view of the system guaranteeing an expected behaviour and a complete knowledge of it, leading them to have a type of 'myopia'.

A holonic architecture allows to bring some correction to the myopia inherent to autonomous and reactive components. To strengthen the systems against uncertainties, specific roles embedding functionalities such as simulations, diagnosis and prediction algorithms can be used.

In order to solve temporal myopia, an example shows that publication of information allows to to reinforce entities' decision making. This information could be used by a coordinator to guide and directly influence behaviour of other entities; or it could be only set accessible by an information responsible (in this case, the component still makes the decision, conserving autonomy). Each of these two alternative are justified according to the context, the events; one major issue is to determine the conditions from which a kind of heterarchy must turn to a kind of hierarchy and vice-versa.

Acknowledgments. This research was financed by the International Campus on Safety and Inter-modality in Transportation, the Nord/Pas-de-Calais Region, the European Community, the French Regional Delegation for Research and Technology, and the French National Centre for Scientific Research. We are grateful for the support of these institutions. A corresponding author is supported by the scholarship program 'Francisco Jos de Caldas - Generación del Bicentenario' of The Administrative Department of Science, Technology and Innovation - COLCIENCIAS and Javeriana University at Bogotá, Colombia, where he is currently an assistant professor.

References

1. Adam, E., Berger, T., Sallez, Y., Trentesaux, D.: An open-control concept for a holonic multiagent system. In: Mařík, V., Strasser, T., Zoitl, A. (eds.) HoloMAS 2009. LNCS, vol. 5696, pp. 145–154. Springer, Heidelberg (2009)
2. Adam, E., Grislin, E., Mandiau, R.: Flexible hierarchical organisation of role based agents. In: Self-Adaptive and Self-Organizing Systems Workshops, pp. 186–191. IEEE Computer Society, Los Alamitos (2008)
3. Cardin, O., Castagna, P.: Using online simulation in holonic manufacturing systems. Engineering Applications of Artificial Intelligence 22(7), 1025–1033 (2009)
4. Heragu, S.S., Graves, R.J., Kim, B.I., Onge St., A.: Intelligent agent based framework for manufacturing systems control. IEEE Transactions on Systems, Man and Cybernetics, Part A 32(5), 560–573 (2002)
5. Johnston, K.: Extending the marketing myopia concept to promote strategic agility. Journal of Strategic Marketing 17(2), 139–148 (2009)
6. Koestler, A.: The Ghost in the Machine. Arkana Books (1969)
7. Leitao, P., Restivo, F.: Adacor: A holonic architecture for agile and adaptive manufacturing control. Computers in Industry 57(2), 121–130 (2006)
8. Marik, V., McFarlane, D.: Industrial adoption of agent-based technologies. IEEE Intelligent Systems 20, 27–35 (2005)
9. Ounnar, F., Pujo, P., Mekaouche, L., Giambiasi, N.: Integration of a flat holonic form in an hla environment. Journal of Intelligent Manufacturing 20, 91–111 (2009)
10. Pujo, P., Broissin, N., Ounnar, F.: Prosis: An isoarchic structure for hms control. Engineering Applications of Artificial Intelligence 22(7), 1034–1045 (2009)
11. Trentesaux, D.: Distributed control of production systems. Eng. Appl. Artif. Intell. 22, 971–978 (2009)
12. Zbib, N., Pach, C., Sallez, Y., Trentesaux, D.: Heterarchical production control in manufacturing systems using the potential fields concept. Journal of Intelligent Manufacturing, 1–22 (2010)

Agent-Based System for Planning and Coordination of Intermodal Transport

Paulina Golinska and Marcin Hajdul

Abstract. The common access to the broadband Internet allows small and medium enterprises (SME) to joindifferent cooperation network. For over decade European Commission in order to fulfill the obligations of sustainable policy promote intermodal transportation. Despite all these efforts the utilization rate of road transportation is still growing reaching last year 76,9%. Intermodal cargo may be carried out in one and the same loading unit or vehicle, on the whole route from the sender to the consignee using the different modes of transport. Planning and coordination of intermodal transportation requires participation of different companies. The main problem still is lack of consistency between processes which are undertaken by various supply chain participants. The aim of this paper is to propose the concept of broader cooperation between intermodal transport process participants based on semantic web and multi-agent system. The application of semantic web and multi-agent system allows to improve the information exchange among different companies and better intermodal transport planning and coordination.

Keywords: multi-agent system, intermodal transportation, sustainability.

1 Introduction – Project Background

Despite of the transitional financial crisis, the European Union countries have reported continuous growth of the carriage of goods by road transport. According to the statistics of the European Commission the share of road transport in total freight (tkm) realized in European countries was 76,9% in 2009. While at the same time the share of rail transport was only 17.6%[1]. The number of roads which has reached its maximum throughput is growing. The effective organization

Paulina Golinska
Poznan University of Technology, Strzelecka 11, 60965 Poznan, Poland
e-mail: {paulina.golinska}@put.poznan.pl

Marcin Hajdul
Institute of Logistics and Warehousing, Estkowskiego 6, 61755 Poznan, Poland
e-mail: {marcin.hajdul}@ilim.poznan.pl

[1] http://epp.eurostat.ec.europa.eu, data for EU 27.

J.M. Corchado et al. (Eds.): Trends in PAAMS, AISC 90, pp. 99–107.
springerlink.com

of transport processes is becoming increasingly more difficult or practically impossible. As the previous studies conducted by authors have proofed, the companies in order to meet quick response requirements deliver the orders of its customers mainly by road.

This approach makes the company improve their processes and thereby increasing the level of customer satisfaction in the short term, but simultaneously may lead to the deterioration of their operations in the longer term. Frequent and rapid deliveries require the involvement of more means of transportation which usually are only partly loaded and this approach increase the traffic leading to congestion and accidents on the road. More over the traffic jams on the congested roads effect in the decrease of the average speed. It results in late deliveries and dissatisfied customers and in the worst casescenario the loss of contracts [3]. The problem of increasing traffic congestion and decreased safety on the road is addressed in a number of the European Commission documents presenting a strategy for the development of the transport and logistics. One of the main ideas, published in the white paper, was to increase the share of alternative modes of transport for carriage of goods by road, particularly rail services. A strong emphasis was also placed on the use of intermodal transport, i.e. those in which the loads are moved by the means of different transport modes [7]. Intermodal cargo istransported in one and the same loading unit or vehicle, on the whole route from the sender to the consignee without the cargo being unloaded itself, using the different modes of transport liketruck, rail e or vessel, barge.

The UE sustainable policy could gain from better planning and coordination of intermodal transport. The higher rate of intermodal transportation utilization benefits in decreasing the traffic congestion, the improvement of road safety and reduction of CO2 emission due to downsizing the number of the only partly loaded tracks. The advantages are clearly stated but still a good information exchange and decision making is missing. A shift in cooperation model is needed among network participants.

The project idea starts from the assumption that through cooperation companies should rationalize their logistics processes, obtain cost savings and reduce empty shipments. At the moment many small and medium sized companies don't activate collaborations. Especially SME's haven't integrated information systems that allow monitoring of data related to the outbound freight flows. This would require significant changes in the transportation process planning and coordination. In order to solve the above mentioned problems a new model ComTra is proposed.

2 Intermodal Transport's Network in Poland

Poland, due to its geographical location, may use virtually all means of transport needed for intermodal system. In practice, however, intermodal transport with the use of inland waterway in Poland does not exist. At present a business practice is combining maritime transport, the rail and road transport. Intermodal transport operations are carried out in intermodal transport units. UTI (frUnités de Transport Intermodal.), which have been codified by the railways of the European countries. These units are: containers large bodywork, car bodies, semi-trailers. However at presents over 98% of goods are carried in containers. An essential element of the intermodal transport system is efficient transport infrastructure. In the case of Poland, this should be good-quality railway lines and container terminals' network, which would allow the handling of logistics units between different means of transportation.

In the last 5 yearsa significant increase in the number of the container terminals has been observed. Over 50% of new terminals were established on the territory of Poland. Unfortunately, despite the significant development of the intermodal infrastructure score, in Poland, the share of intermodal transport in rail transport is estimated on the 2.5%-3%. Currently, most of the containerstransport from polish seaports is done by road (about 80%). In 2008 four ports (Gdansk, Gdynia Szczecin and Świnoujście) unloaded 859 183 TEU, therefore more about 20% than the total quantity of containers transported by intermodal transport in Poland. Similarly, the situation looks with the carriage of containers to Poland from the main European ports, such as Rotterdam, Hamburg or Bremerhaven. In 2008 nearly 650 000 TEU in the import or export from Poland was handled by the road transport. Therefore there is a significant market potential for the development of intermodal transport in the future(up to10-15% of cargo in total).

3 Agent-Based System

The above described intermodal transport network requires close cooperation between a big number of independent entities.Often problem is a lack of the coordination and integration between the information systems working within and between network members. It can be stated that following characteristics are typical for intermodal transport network:data, control and resources are inherently distributed; the system is regarded as society of autonomous cooperating entities; the components must interact with other to reach whole system goals.

Agent technologies can be used to facilitate and control the scope, the time, and the frequency of information sharing based on specific arrangement of each partners. The distributed multi-agent system is suitable to perform inter-organizational planning of intermodal transport processes.Agent technology is the preferable technology for enabling a flexible and dynamic coordination of spatially distributed entities [8]. The mentioned above characteristics suit well to requirement for planning and coordination of intermodal transport process. In an agent-enabled environment, the multiple transactions and exchanges of information that drive material flow can be managed by software agents that maintain visibility across the network [8]. Multi-agent systems perfectly suit the demands for global flexibility, cooperation and at the same time, local autonomy.

Thecharacteristic of agent technology suits well to perform planning and coordination in an intermodal network environment where we can observe:heterogeneous environment, open and dynamic structure, interoperability of components, need for scalability.

The effectiveness of the transport's network might be measured by its ability to provide customer with due-to-date deliveries of final goods at optimal cost and in environmental friendly way. In order to meet this goal the effective and integrated management of material and information flow at a tactical and operational level is of crucial importance. Agent-based system architecture provides an interesting perspective for integration of the supply chain due the fact that they are able to generate, process, filter, broadcast, and correlate information for real-time coordination and decision making.

In this paper an agent-based system is treated as a multi-agent system that's acts as a support tool to company existing business information system (for example ERP) and utilized their databases [2].Authors apply Jennings and Wooldridge's definition [4]: "an agent is a computer system situated in some environment, and that is capable of autonomous action in this environment in order to meet its design objectives. Agents have the following characteristic [4]: reactive, pro-activeness, social ability.

The ComTramodel is a hybrid solution elaborated based on agent technology and semantic web concept including such elements, as XML, Resource description Framework (RDF), and ontologies. Semantic web is an initiative that aims to introduce standard description of the content in Internet in order to allow computer programs(eg. agents) processing of data and information appropriate to their purpose and meaning [1].

The main feature of semantic web is its interoperability. It is understood as an ability of different information systems to cooperate, safely exchange data with predefined structure,as well as thefurther mutual usage of this data in order to create information. Communication between IT systems is possible independently of programming languages, platforms and operations systems they use, and applied information exchange standard [5].All the above mentionedfeatures make the semantic web suitable for application in the e-transportation (virtual) networks.

4 ComTra Model

4.1 ComTra–Model Process Layer

In the created the model there are agents: manufacturing companies, transport service providers, intermodal transport processes coordinators. Model ComTra also defines the roles played by participants (agents), the dependencies between them (inside the virtual supply chain) and between cooperating companies, a region in which they are active. The methodology takes into account the strong relationship between scale micro (the companies), and the macro scale (the region in which the companies perform their operations).

In figure 2 the reference model for planning and coordination of intermodal transport process is presented. It is based on the sustainable development framework.

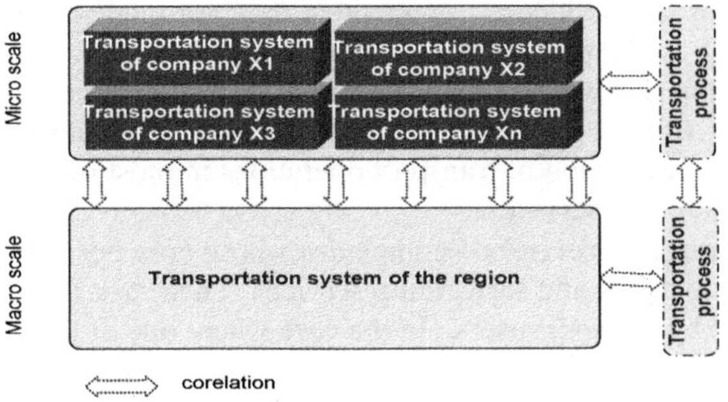

Fig. 1 The interdependencies between micro and macro scale transportation systems

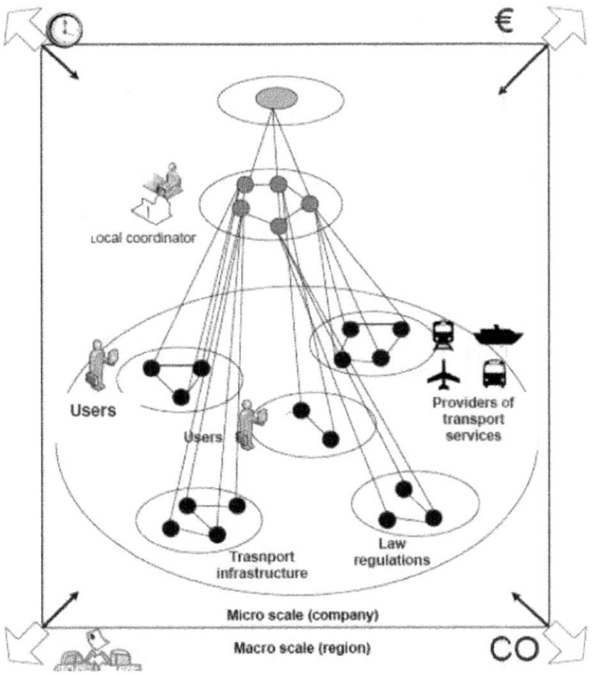

Fig. 2 The reference model for intermodal transport planning and coordination

Cooperating entities shall exchange information electronically, using dedicated electronic platforms. In the ComTramodel three groups of playersare defined (see Figure 3):

- Users of transport services – companies that are engaged in the production and/or selling of the products. Transport is not their main source of business. These companies may have their own means of transport or they cooperate with the providers of transport services. They issue the demand for transportation. They order the execution of transport operations in most of the cases by issuing order to providers of services.
- Provider of transport services – companies whose core business is the provision of transport, logistics and forwarding services. Their task is the coordination of orders issued by transport users. In the case where one of the cooperating firms has its own means of transport and is able to provide transport services for the other transport users, then it performs also the role of transport services providers.
- Coordinator – represents the transport users, deals with the coordination of transport processes (e.g. analysis of the possible aggregation of the transport orders issued by the different users, price negotiations, the choice of modes of transport), cooperation with suppliers of transport services, monitoring of the aggregated processes of transport, the carrying out of analyses in the micro and macro. The role of the coordinator may be performed in several ways. Coordinator might by financed by cooperation companies in the network, when a coordinator is their employee. The coordinator may be an independent entity supporting coordination of transport processes, which charges the users on the basis of the number of orders being executed. Finally, this may be the shipping company, acting exclusively for companies that have decided to collaborate in the organization of the transport processes.

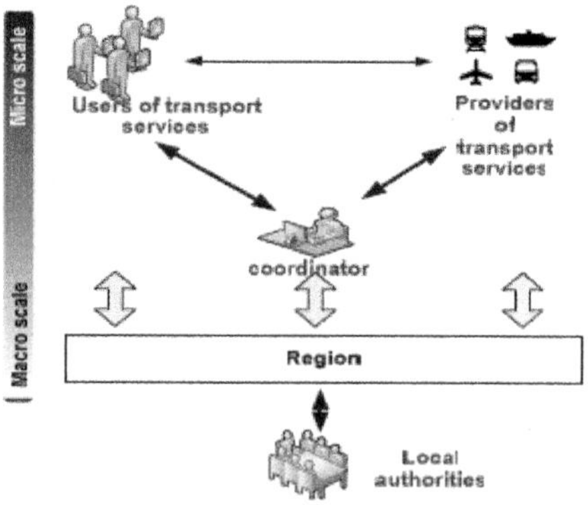

Fig. 3 Players of the ComTra model

The actionstaken by each of the above mentioned players have a direct impact on the transport system of the region in which they operate. In addition, each of the predefined entities is also vulnerable to the impact of the environment (e.g. decisions of public administration), in this case, the transport system of the region.

4.2 ComTra–Model Information Layer

As mentioned before the participants of intermodal transport network should used web-based platform for exchange of the information. In the ComTra model the stationary management agent [6] (that has in command the other agents) is implemented. In order to gain information the Coordination Agent (CA) delegates collecting of predefined information scope about potential business partners to group of mobile agents. CA represents the particular interests of cooperating companies and it is responsible for cooperation, communication, negotiation. On the basis of collected information the CA makes decisions regarding follow-up actions. Coordination Agent is able to manage complex business processes. It cooperates with groups of agents and creates multi-agent system, as followed:

- Users Agent (UA) – it is responsible for issuing the demand and the definition of criteria for different cooperation scenario. It is able to inform the Verifying Agent about semantic inconsistency of found information.
- TransportationProviders Agent (TPA) – represents companies which offer particular transport services
- Informing Agent(IA) – it is responsible for sending information to CA about new attractive offers available, as well as for informing clients agents and transport providers agent about the possibility of starting cooperation with CA after fulfillment of defined criteria.
- Verifying Agent (VA) – it verifies ontologies and acts together with tool for semantic verification of information placed in the web by trading partners.

Agents representing particular companies cooperate together and negotiate cooperation conditions in order to reach common goal and at the same time to fulfill the outlines of delegating institutions. It should be mentioned that agents representing different, sometimes competing companies can build up temporary alliances in order to reduce the cost of logistics services and environmental impact.

The communication between agents and user is asynchronic, so the confirmation of message by receiver is not required in order to continue the process execution [5]. Agents and users are informed automatically about any new event e.g.new offer arrival.

The access to the semantic web does not require application of any specialized IT systems. Information can be process in format that is readable and understandable for computer and any user. This solution gives SMEs a better possibility to capture information and exchange them withinparticular enterprises'network, as well as it helps to build up business relations [5].

In figure 4 the schema for information exchange between network participants is described.

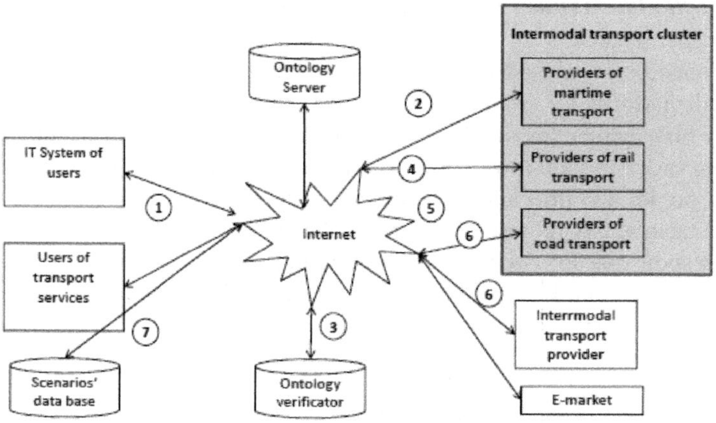

Fig. 4 Model of multi-agent system based on the semantic web. (source modified from [5])

Application of the agent-based system and semantic web allows the improvement of the intermodal transport process. The procedure below presents the process of planning intermodal transport by agents, following the defined scenario (see Fig. 4):

1. Users of transport services order their agents (UA) issuing demand for transport services and a choice of carrier for shipment of goods according to sustainability and economic optimization criteria (intermodal transport mode).
2. UA through hyperlinks finds ontology defining key words. Then it communicates with TPA representing potential service providers and collects their offers and passes them to CA for analysis.
3. In case when UA has doubts about the content of the offer, it could command Verifying Agent (VA) to verify ontology and to contact its issuer.
4. Before offer choice CA questions providers agent TPA whether the offer is still valid. If not, then the whole procedure needs to be repeated. If the offer is still valid, then the appropriate resources are booked. CA can book multiple carriers, especially in case when individual carrier is not able to provide intermodal transport service. Particular TPA can create intermodal transport cluster.
5. After the offer is chosen CA informs agents representing business partners UA and TPA and starts negotiations regarding number of routes, delivery dates, prices etc.
6. If the negotiations are successful, then the cooperation starts and agents sign the commission of service purchase on behalf of companies they represent.
7. When the transaction is finished, CA stores the scenario in its database in order to simplify the negotiation process in the future.

5 Conclusions

Almost unlimited access to the Internet makes possible cooperation in the area of logistics process not only between big company but also SMEs. The lack of consistency of business process performed by particular entities and the variety IT systems used by companies, cause problems with automatic partners networking. The presented by authorsComTramodel provides framework for planning and coordination of intermodal transport in micro (company) and macro scale (region). Proposed solution enables efficient creation and automatic collection of data about companies and their resources published in the Internet. Agents representing particular companies coordinate and establish cooperation's conditions, in order to reach common goal of cost effectiveness as well as the sustainable development goals.

Acknowledgments. Presentedsolution has been partly developed within own research project N509 398536 – "Intermodal logistics network in Poland – model solution and condition for implementation", funded by Ministry of Science and Higher Education.

References

1. Berners-Lee, T., Hendler, J., Lassila, O.: A New Form of Web Content that is Meaningful to Computers will Unleash a Revolution of New Possibilities. Scientific American 284(5) (2001)
2. Golinska, P., Fertsch, M., Marx Gomez, J., Oleskow, J.: The concept of Closed-loop Supply Chain Integration Through Agent-based System. In: Marx Gomez, J., et al. (eds.) Information Technologies in Environmental Engineering, pp. 189–202. Springer, Heidelberg (2007)
3. Hajdul, M.: Racjonalizacja systemów transportowych przedsiębiorstw szansą nazmniejszenie ruchu pojazdów ciężarowych w Wielkopolsce, Logistyka (April 2007)
4. Jennings, N.R., Wooldridge, M.J.: Applications of Intelligent Agents. In: Agent Technology: Foundations, Applications and Markets. Springer, Heidelberg (1998)
5. Kawa, A., Pawlewski, P., Golinska, P., Hajdul, M.: Cooperative purchasing of logistics services among manufacturing companies based on semantic web and multi-agents system. In: Demazeau, Y., et al. (eds.) Trends in PAAMS. AISC, vol. 71, pp. 249–256. Springer, Heidelberg (2010)
6. Wieczerzycki, W.: Polymorphic Agent Clusters – the Concept to Design Multi-Agent Environments Supporting Business Activities. In: Mařík, V., William Brennan, R., Pěchouček, M. (eds.) HoloMAS 2005. LNCS (LNAI), vol. 3593, pp. 233–245. Springer, Heidelberg (2005)
7. Rydzkowski, W., Wojewódzka-Król, K. (red.): Transport, Wydawnictwo Naukowe PWN, Warszawa (2009)
8. Villa, A.: Emerging trends in large-scale supply chain management. International Journal of Production Research 40(15), 3487–3498 (2002)

Boosting Bitext Compression

Joaquín Adiego, Miguel A. Martínez-Prieto, Javier E. Hoyos-Torío,
and Felipe Sánchez-Martínez

Abstract. Bilingual parallel corpora, also know as bitexts, convey the same information in two different languages. This implies that when modelling bitexts one can take advantage of the fact that there exists a relation between both texts; the text alignment task allow to establish such relationship. In this paper we propose different approaches that use words and *biwords* (pairs made of two words, each one from a different text) as representation symbolic units. The properties of these approaches are analysed from a statistical point of view and tested as a preprocessing step to general purpose compressors. The results obtained suggest interesting conclusions concerning the use of both words and biwords. When encoded models are used as compression boosters we achieve compression ratios improving state-of-the-art compressors up to 6.5 percentage points, being up to 40% faster.

Keywords: Compression Boosting, Bitext Compression.

1 Introduction

As a consequence of the globalisation and the existence of countries and supra national entities embracing regions with different languages the amount of texts that are stored together with their translation into different languages has dramatically increased. A text placed together with its translation into other languages is referred to as a *multilingual parallel corpus*; a *bilingual parallel corpus*, also know as *bitext*, is a parallel corpus made of two texts.

Joaquín Adiego · Miguel A. Martínez-Prieto · Javier E. Hoyos-Torío
Dpto. de Informática, Universidad de Valladolid, Spain
e-mail: {jadiego,migumar2}@infor.uva.es, javierht@gmail.com

Felipe Sánchez-Martínez
Dpto. de Llenguatges i Sistemes Informàtics, Universitat d'Alacant, Spain
e-mail: fsanchez@dlsi.ua.es

J.M. Corchado et al. (Eds.): Trends in PAAMS, AISC 90, pp. 109–116.
springerlink.com

Bitexts are, as stated by Melamed [10], "one of the richest sources of linguistic knowledge because the translation of a text into another language can be viewed as a detailed annotation of what that text means".

The use of bitext by human language technology applications has grown in parallel with the availability of large collections of bitexts. These applications may use compression techniques to solve storage problems and improve access time in storing and processing [13] by trading disk transfer operations for processor operations. In addition, the use of compression techniques reduces transmission times, which increases the efficiency of communications.

Next section explains the main concepts related to the compression of bitexts and gives a brief description of related work. Section 3 covers some modelling policies to represent bitexts, Section 4 then explains a new modelling technique and gives several statistics. Section 5 discusses the results achieved when compressing different bitext corpora. The paper ends with some conclusions.

2 Related Work

A *bitext* consists of two texts that are mutual translations. A considerable part of the information contained in a bitext is highly redundant because the same semantic content is represented in two different ways. Storage and transmission costs are thus unnecessarily increased by storing bilingual versions of the same content. Ideally, this fact could be avoided if perfect machine translation systems were available, since only one version of the text would be needed to produce its translation [11].

A bitext in which the translation relationship between the words in one text (left) and the words in the other text (right) has been established is usually referred to as a *word-aligned bitext*. A way to benefit from the fact that both texts in a bitext (say, L and R) are mutual translations is to know for each segment in L which is its counterpart in R. The *word alignment* task [12] connects words in text L with those in R. The result is a bigraph for the words in L and the words in R with an arc between word $l \in L$ and word $r \in R$ if and only if they are mutual translations. Word alignment is done after *sentence alignment* [7] which, analogously, identifies pairs of sentences that are mutual translations.

Nevill-Manning and Bell [11] explore how traditional text-compression methods can be extended to the compression of bitexts. Their model is based on two types of relations: exact correspondences between two words and synonymy relationships between the words in both texts (given by a thesaurus).

In [4], *text alignment* is proposed as the basis for bilingual text compression. The alignment-based algorithm requires, in addition to the left (L) and right (R) texts, the existence of word- and phrase-level alignments, the lemmatised forms of L and R, a lemmata dictionary of words in L, a variant dictionary with the lemmata of all words in R and a bilingual glossary. Slight improvements are reported [4] with respect to classical compression algorithms such as BZIP2 or word-oriented Huffman.

3 Modelling Bitexts

To preserve the connection established in the word alignment process which, in turn, allow us to recognise which text segments are mutual translations, a bitext may be represented as a sequence of word pairs (*biwords*) where the first element corresponds to a word in the left text and the second element refers to its counterpart in the right text. An *empty word* (ϵ) is assumed as the counterpart word if a word in one text is not aligned with any word in the other text, or if its alignment was discarded in order to be able to restore the original texts when decompressing. For instance, the biword representation of the Spanish–English bitext *(la casa donde vivimos, the house where we live)* is *(la,the), (casa,house), (donde,where), (ϵ,we) (vivimos,live).*

Using a word-based model enhanced with the empty word symbol may be seen as a reasonable choice to represent a bitext. This implies to perform a pair-to-pair parsing and to identify each word as an independent symbol. These symbols are then encoded in such a way that the codewords assigned to the words in the pair are contiguous in the compressed text. Codeword assignment can be made in two different ways. On one hand, a single vocabulary can be used to represent all the words in the bitext (1V); this is a good choice to compress bitexts made of closely-related languages which share many words. On the other hand, two different vocabularies, one for the words in the left text and another for the words in the right text, may be used. This last method (2V) requires to represent the empty word in both vocabularies. Although these two approaches allow to get compressed bitext representations, they are not effective because the same information is represented twice, enlarging the bitext representation.

To represent the two texts that comprise a bitext on a single model, the concept of *biword* was introduced by Martínez-Prieto et al. [9]. The use of biwords allows to represent with a single symbol two words with high mutual information [5]. The main drawback of using biwords is that larger dictionaries are needed; however, this is not an obstacle to achieve a large spatial saving when a bitext is compressed by using biwords as the symbols to compress.

A biword-based scheme, called 2LCAB, is proposed in [1]. 2LCAB builds a two-level dictionary in which word and biword representations are stored. The aligned bitext representation is pair-to-pair parsed and each word in the pair is then represented in the first-level vocabulary corresponding to its language. Finally, each pair of words (biwords) is represented as a single symbol in the biword vocabulary. ETDC [3] is used to encode both the words in isolation and the biwords. In 2LCAB words in the first-level dictionaries are ranked in accordance with the number of different biwords in which they appear.

Biwords in the second-level vocabulary are ranked in accordance with their frequency in the bitext. However, no biword strings are used when the biword vocabulary is stored with the compressed bitext, instead the concatenation of the codewords of the two words present in the biword are used. 2LCAB allows the bitext to be processed (searched and retrieved) in its compressed form.

4 First-Order Model on Translations Relationships

Bearing in mind Melamed's [10] affirmation which states that the translation of a text into another language can be viewed as a detailed annotation of what that text means, we propose a new model (TRC_1) for bitexts in agreement with this affirmation. The idea is to represent the words in the right text with respect to the preceding word in the left text, thus, a first-order model based on translation relationships is proposed. In a dictionary we represent all the different words in the left text ranked in accordance with their frequency in the bitext; an independent "translation" dictionary is then associated to each left word in this (left) dictionary. This second dictionary stores all the different words in the right text paired in a biword with the corresponding word in the left dictionary. Words in associated "translation" dictionaries are ranked in accordance with the number of different biwords in which they appear. With the exception of the empty word, each word in a texts is related to a small number of words in the other text and, therefore, one byte should be sufficient to code the right word in each biwords. Figure 1 illustrates the dictionaries used by TRC_1.

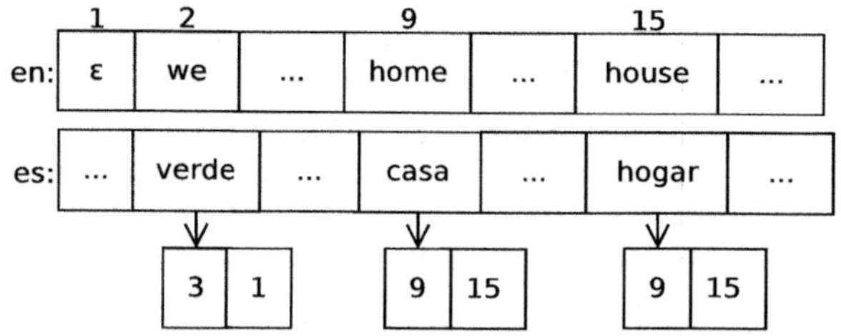

Fig. 1 Dictionaries used by TRC_1. The translation dictionaries link the words in the left dictionary (es) with those in the right dictionary (en) by storing the corresponding indexes.

While the possible gain of this method is clear when it is used as a preprocessing step to another compressors, as a side effect it requires to encode two symbols (left and right words) instead of just one (biword). It is therefore worth exploring if this permits to achieve better compression ratios since, as

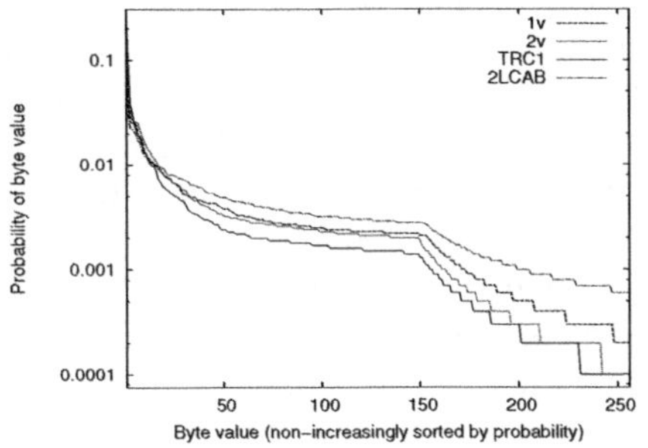

Fig. 2 Relationship between byte values (non-increasingly sorted by probability) from a Spanish–English bitext

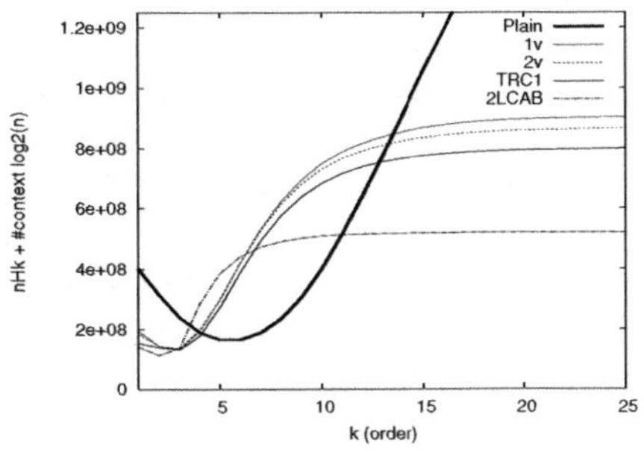

Fig. 3 Values $nH_k + \#\text{contexts} \times \log_2(n)$ for plain and different models codified with ETDC from a Spanish–English bitext

a previous step to another compressor, coding biwords with a single symbol results in larger dictionaries and the redundancy loss in the encoded stream may be a handicap for the compressor.

Figure 2 plots all byte frequencies when Spanish–English bitexts are modelled using the 1V, 2V, 2LCAB and TRC_1 aforementioned approaches and their symbols are encoded via ETDC [3]. As can be seen, the byte frequency distribution is very skewed, which suggests compressing this byte stream with a bit-oriented technique. As the order grows, compression improves because the model captures larger correlations between consecutive characters in the text but, unfortunately, the number of different contexts becomes unmanageable. The average length of a word is close to 5 bytes in English texts [2] but the variance is relatively high and is raised if attention is paid to the distance between two consecutive words. A text encoded with ETDC is obviously advantageous because the average length of a word is around 2 bytes with low variance. A k-order modeller can therefore capture the correlation between consecutive words with a much smaller k, or capture longer correlations with a given k [6]. Figure 3 provides a realistic estimation of the size of the compressed bitext achievable by a k-th order compressor. It displays the value nH_k as a function of the number of contexts, where n is the size of the text. Notice that these values are corrected by penalising each new context with $\log(n)$ bits [6].

5 Evaluation

Experiments have been performed in order to test, on one hand, the efficiency of different approaches to compress bitexts and, on the other hand, the effect of the bitext size on the compression ratio. We used the following corpora: (i) a Spanish–Catalan (`es-ca`) bitext from *El Periódico de Catalunya*,[1] a daily newspaper published both in Catalan and Spanish; (ii) a Spanish–Galician (`es-gl`) bitext from *Diario Oficial de Galicia*,[2] the bulletin of the Government of Galicia, published both in Galician and Spanish; and (iii) bitexts for Spanish–English (`es-en`), French–English (`fr-en`), and German–English (`de-en`) from the *European Parliament Proceedings Parallel Corpus* [8].

We created bitexts of different sizes for each language pair in such a way that larger bitexts contain smaller ones. To compute the word alignments we used the largest bitext for each language pair and the GIZA++ statistical aligner [12].[3] An isolated AMD Athlon Dual Core processor at 2 GHz with 2 GB of RAM and running Debian 4 Etch was used. We used a g++-4.1.2 compiler with full optimisation; time results measure CPU user time.

We have tested the two word-based approaches described in Section 3: 1V uses a single vocabulary in which all words in the bitext are represented, whereas 2V considers two vocabularies to store independent representations of the words in each text. Both approaches use the *encode* function defined by ETDC [3] for codewords assignment. We do not compare with the approach proposed by Conley and Klein [4] because we have not found any implementation to test it with the bitext collection mentioned above. Anyway, Conley and Klein compare their TRANS approach with GZIP and BZIP2 and they conclude that TRANS is slightly better that BZIP2 (an improvement of 1% is achieved). However, the authors do not consider the size of the auxiliary files that TRANS requires to decompress the bitext.

Table 1 shows the compression ratios achieved when 1V, 2V, 2LCAB and TRC_1 are used as compression boosters; in this case, they are used as a preprocessing step to the well-know GZIP, BZIP2 and PPMDI compressors. Results show that 1V+GZIP, 2V+GZIP, 2LCAB+GZIP and TRC_1+GZIP obtain the worst compression ratios in their category. However, for large bitexts consisting of closely-related languages they improve PPMDI up to 4.5 percentage points (they can be directly compared), and BZIP2 up to 6.3 percentage points. When the bitexts consist of less-related languages the improvement is reduced to 2.1 percentage points for PPMDI and to 3.8 percentage points for BZIP2. 1V+PPMDI, 2V+PPMDI, 2LCAB+PPMDI and TRC_1+PPMDI obtain the best compression ratios, improving PPMDI up to 4.0 percentage points.

Compression ratios obtained when 2LCAB and TRC_1 are used as a preprocessing step are very similar, although TRC_1 approximately encodes twice as symbols as 2LCAB. In fact, TRC_1 is better than 2LCAB for small bitexts and for bitexts consisting of closely-related languages.

Table 2 shows compression and decompression times, respectively; word alignment times are note taken into account. Due to lack of space we only show 2LCAB and TRC_1 as boosters times, however all the times obtained by using 1V and 2V as boosters are similar to those of TRC_1. Results show

[1] `http://www.elperiodico.com`

[2] `http://www.xunta.es/diario-oficial`

[3] `http://code.google.com/p/giza-pp/`

Table 1 Compression ratios achieved by 1V, 2V, 2LCAB and TRC_1 when they are used as compression boosters

	MB	1V			2V			2LCAB			TRC_1		
		GZIP	BZIP2	PPMDI	GZIP	BZIP2	PPMDI	GZIP	BZIP2	PPMDI	GZIP	BZIP2	PPMDI
es-gl	1	12.43%	9.84%	8.72%	14.16%	11.38%	10.12%	12.88%	11.61%	10.82%	13.27%	10.76%	9.59%
	10	10.77%	7.42%	6.88%	11.61%	8.09%	7.47%	9.16%	7.35%	6.90%	9.96%	7.24%	6.52%
es-ca	1	28.59%	23.86%	21.82%	29.79%	25.08%	23.24%	28.63%	26.23%	25.28%	28.13%	24.24%	22.34%
	10	23.42%	18.21%	17.81%	23.94%	18.69%	18.20%	19.61%	17.46%	16.65%	20.18%	16.98%	15.76%
	100	21.34%	16.52%	16.24%	21.53%	16.65%	16.30%	15.67%	13.87%	13.25%	16.87%	14.07%	13.11%
es-en	1	26.36%	23.43%	21.30%	25.52%	23.19%	21.00%	26.73%	25.43%	24.09%	26.22%	24.33%	22.12%
	10	22.94%	19.53%	18.34%	22.06%	19.29%	17.97%	20.67%	19.20%	18.29%	21.22%	19.31%	17.74%
	100	21.27%	17.97%	16.95%	20.43%	17.70%	16.58%	17.36%	16.11%	15.40%	18.68%	16.91%	15.56%
fr-en	1	25.69%	22.82%	20.71%	25.06%	22.73%	20.59%	26.18%	24.90%	23.59%	25.76%	23.82%	21.66%
	10	22.57%	19.17%	17.99%	21.79%	19.02%	17.71%	20.34%	18.83%	17.96%	20.93%	18.95%	17.43%
	100	21.31%	18.09%	17.03%	20.49%	17.84%	16.67%	17.54%	16.25%	15.56%	18.83%	17.03%	15.69%
de-en	1	26.93%	24.08%	21.88%	25.94%	23.71%	21.49%	27.38%	25.95%	24.61%	26.92%	24.92%	22.60%
	10	23.44%	20.07%	18.88%	22.49%	19.85%	18.45%	21.32%	19.76%	18.84%	21.90%	19.87%	18.22%
	100	21.90%	18.90%	17.76%	21.00%	18.61%	17.34%	18.30%	17.07%	16.35%	19.61%	17.86%	16.42%

Table 2 Compression/Decompression times achieved with different bitext corpora. Word alignment times are not taken into account.

MB	GZIP	BZIP2	PPMDI	2LCAB			TRC_1		
				GZIP	BZIP2	PPMDI	GZIP	BZIP2	PPMDI
es-ca									
1	0.23/0.04	0.77/0.25	0.91/0.98	0.37/0.09	0.55/0.17	1.09/1.03	0.50/0.10	0.65/0.19	1.15/1.04
10	1.94/0.25	4.58/1.94	5.31/5.84	2.70/0.44	3.59/1.34	6.22/5.50	3.30/0.47	4.18/1.58	6.85/5.90
100	9.60/1.88	35.34/12.19	38.83/41.46	10.26/2.53	17.43/6.92	31.36/30.22	15.67/3.61	22.32/8.80	37.14/33.19
es-en									
1	0.23/0.03	0.73/0.25	0.87/0.95	0.33/0.06	0.50/0.16	0.97/0.89	0.43/0.06	0.57/0.18	0.97/0.86
10	1.85/0.18	4.40/1.76	5.43/5.56	2.27/0.41	3.33/1.29	6.06/5.53	3.10/0.48	4.19/1.53	6.02/5.45
100	12.11/1.90	41.68/14.70	46.98/50.22	12.06/2.84	19.94/7.63	37.59/35.67	17.72/3.25	24.95/8.96	42.13/36.41

that the techniques we propose are very fast at compression and, mainly, at decompression. This is explained by the fact that both BZIP2 and PPMDI are very slow and when they compress the transformed text they are actually compressing 30% of the plain text. As a result, not only 2LCAB+BZIP2 and 2LCAB+PPMDI compress more than BZIP2 and PPMDI respectively, but also they are up to 40% faster.

6 Conclusions

We have shown and analysed some modelling proposals that use words (1V, 2V and TRC_1) and biwords (2LCAB) as symbols. When a bitext is modelled with the aforementioned techniques and encoded with a well-know byte-oriented code, as ETDC, acceptable compression ratios are obtained. In addition, they can be seen as a transformation of the bitext that boost general purpose compressors because they transforms a bitext in a shorter byte sequence (19-50% of the original) that can still be compressed. On the other hand, we have shown that the compression ratios reported are coherent with the statistics of the representation model and the previous analytical study.

When encoded models are used as a preprocessing step to general-purpose compressors, the experiments show that they improve the compression ratio as well as their performance in both compression and decompression yielding an attractive space/efficiency trade-off.

Acknowledgements. Work supported by the Spanish Government through projects TIN2009-14009-C02-01 and TIN2009-14009-C02-02; and by the Millennium Institute for Cell Dynamics and Biotechnology (ICDB) (Grant ICM P05 -001-F).

References

[1] Adiego, J., Brisaboa, N.R., Martínez-Prieto, M.A., Sánchez-Martínez, F.: A two-level structure for compressing aligned bitexts. In: Karlgren, J., Tarhio, J., Hyyrö, H. (eds.) SPIRE 2009. LNCS, vol. 5721, pp. 114–121. Springer, Heidelberg (2009)

[2] Bell, T.C., Cleary, J.G., Witten, I.H.: Text Compression. Prentice Hall, Englewood Cliffs (1990)

[3] Brisaboa, N.R., Fariña, A., Navarro, G., Paramá, J.R.: Lightweight natural language text compression. Inf. Retr. 10(1), 1–33 (2007)

[4] Conley, E.S., Klein, S.T.: Using alignment for multilingual text compression. Int. J. Found Comput. Sci. 19(1), 89–101 (2008)

[5] Cover, T.M., Thomas, J.A.: Elements of Information Theory. Wiley, Chichester (1991)

[6] Fariña, A., Navarro, G., Paramá, J.: Word-based statistical compressors as natural language compression boosters. In: Proc. 18th Data Compression Conference (DCC), pp. 162–171 (2008)

[7] Gale, W.A., Church, K.W.: A program for aligning sentences in bilingual corpora. Comput. Linguist. 19(1), 75–102 (1993)

[8] Koehn, P.: Europarl: A parallel corpus for statistical machine translation. In: Proceedings of the Tenth Machine Translation Summit, Phuket, Thailand, pp. 79–86 (2005), `http://www.statmt.org/europarl/`

[9] Martínez-Prieto, M.A., Adiego, J., Sánchez-Martínez, F., de la Fuente, P., Carrasco, R.C.: On the use of word alignments to enhance bitext compression. In: Data Compression Conference, p. 459 (2009)

[10] Melamed, I.D.: Emplirical methods for exploting parallel texts. MIT Press, Cambridge (2001)

[11] Nevill-Manning, C.G., Bell, T.C.: Compression of parallel texts. Inf. Process. Manage. 28(6), 781–794 (1992)

[12] Och, F.J., Ney, H.: A systematic comparison of various statistical alignment models. Comput. Linguist. 29(1), 19–51 (2003)

[13] Ziviani, N., Moura, E., Navarro, G., Baeza-Yates, R.: Compression: A key for next-generation text retrieval systems. IEEE Computer 33(11), 37–44 (2000)

Web Document Duplicate Detection Using Fuzzy Hashing

Carlos G. Figuerola, Raquel Gómez Díaz, José L. Alonso Berrocal, and Angel F. Zazo Rodríguez

Abstract. The web is the largest repository of documents available and, for retrieval for various purposes, we must use crawlers to navigate autonomously, to select documents and processing them according to the objectives pursued. However, we can see, even intuitively, that are obtained more or less abundant replications of a significant number of documents. The detection of these duplicates is important because it allows to lighten databases and improve the efficiency of information retrieval engines, but also improve the precision of cybermetric analysis, web mining studies, etc. Hash standard techniques used to detect these duplicates only detect exact duplicates, at the bit level. However, many of the duplicates found in the real world are not exactly alike. For example, we can find web pages with the same content, but with different headers or meta tags, or viewed with style sheets different. A frequent case is that of the same document but in different formats; in these cases we will have completely different documents at binary level. The obvious solution is to compare plain text conversions of all these formats, but these conversions are never identical, because of the different treatments of the converters on various formatting elements (treatment of textual characters, diacritics, spacing, paragraphs ...). In this work we introduce the possibility of using what is known as fuzzy-hashing. The idea is to produce fingerprints of files (or documents, etc..). This way, a comparison between two fingerprints could give us an estimate of the closeness or distance between two files, documents, etc. Based on the concept of "rolling hash", the fuzzy hashing has been used successfully in computer security tasks, such as identifying malware, spam, virus scanning, etc. We have added capabilities of fuzzy hashing to a slight crawler and have made several tests in a heterogeneous network domain, consisting of multiple servers with different software, static and dynamic pages, etc.. These tests allowed us to measure similarity thresholds and to obtain useful data about the quantity and distribution of duplicate documents on web servers.

Carlos G. Figuerola · Raquel Gómez Díaz · José L. Alonso Berrocal ·
Angel F. Zazo Rodríguez
University of Salamanca, C/ Francisco de Vitoria 6-16, 37008 Salamanca, Spain
e-mail: {figue,rgomez,berrocal,zazo}@usal.es

J.M. Corchado et al. (Eds.): Trends in PAAMS, AISC 90, pp. 117–125.
springerlink.com

Keywords: Web Crawling, Fuzzy Hashing, Document Duplicate Detection, Information Retrieval.

1 Introduction

The web is the largest repository of documents available. Even in a purely intuitive way is easy to find abundant information replicated in any tour of the web. Bharat and Broder [2] cite studies of the second half of the 90 in which it is shown that over 30% of web pages explored by crawlers are duplicates. Although in their study identified only 9.4% as duplicate pages, it is also true that they were focused only in duplicates by mirroring. In any case, these are important quantities which raises suspicions of high level of replication of web content for various reasons. Duplication of documents hinders the work of the crawlers and slows down the indexing of documents, is an added difficulty for users who need to review the documents retrieved after a search. Chowdhury and others have noted, moreover, the problem of duplication of documents and their impact not only on the speed of retrieval, but also in the calculation of weights of terms based on their frequencies [4]. In many cases, however, is not exact duplicates, but near duplicates: documents that have virtually the same content, but differ on some details. Yerra and Ng [19] quote some cases, by way of example, of documents in different versions, small documents assembled in a large document, and the reverse: split large documents into smaller parts, different views of the same content ... To this we can add a few more causes: specific headers added by web servers or web applications to identical documents, visitors counters, documents that include the current date, their own URL, relative paths, etc.

Because that, it is important to have techniques and tools to detect duplicate and near duplicate documents, this need arises in various contexts, and which concerns us: that of information retrieval on the web. This article shows some preliminary results in implementing in this type of tasks of techniques known as *fuzzy hashing*. This paper is organized as follows: Section 2 gives a brief review of the most common techniques for measuring the resemblance or similarity between documents, and its application in detecting near duplicate documents. Section 3 describes the *fuzzy hashing* or *Context Triggered Hasshing Piece (CTPH)*. Section 4 describes the experimental tests carried out and discuss its results. Finally we present some conclusions and future work lines.

2 Measuring Similarity between Documents

There are different systems to estimate the resemblance or similarity between two documents. Several of them stem from the calculation of similarity between vectors, apply where the documents are represented by vectors: cosine Dice, Jaccard et al. [16] But they are not useful in this case, as they were designed to estimate semantic similarity, the vector model itself which supporting its use part of the independence between terms, without taking into account the position relative to each other. And in any case, their application has processing requirements inapplicable in this context [3].

Other systems derived from what is known as edit distance, ie the number of operations required to transform a string in another [14]. This includes measures of similarity between sequences of characters such as Hamming [8], Lenvensthein [12], [15] or Damerau [5], among several others. Their main problem is that they are designed for short sequences of characters, one of its most common uses, for example, is the correction of typographical errors by suggesting words close. Applied to very long strings, such as full papers, are inapplicable because they need too long time.

The most used solution to detect duplicate documents is *hashing*, ie obtaining a criptographyc fingerprint wich, because the algorithms used, is small, but fairly reliably ensures that there are not two equal hashes for documents are different. The hashing has diverse applications, from facilitating quick searches os data structures wich use a hash as a key or reference of the stored data, to use encryption, for example, digital signature applications or as a guarantee of no alteration of a document. Some of the most popular are MD5 or SHA, MD5 is often used in free software downloads for checking the integrity of the original files. It is easy to find utilities in the operating systems and programming languages that allow to obtain hashes by these algorithms.

However, conventional hashing, which works very well for the purposes for which it was conceived, has disadvantages in the detection of duplicate doumentos. In fact, the smallest difference between two documents, even at the bit level, produces hashes completely different, being this difference not proportional to the real difference between such documents. Thus, an algorithm as such MD5 certifies pretty sure that two documents are identical, but is not valid to detect documents that, although containing small differences, could be considered equal in practice.

3 Fuzzy Hashing

The *Fuzzy hashing* o *Context Triggered Piece Hashing (CTPH)* is based in works by A. Tridgell [18], who, after using hash techniques in the development of *rsync* (a popular GNU application for the incremental transfer of files, `http://rsync.samba.org/`) applied *fuzzy hash* in another application (*spamsum*) dedicated to detecting spam e-mail messages.

The *fuzzy hash* is based on what is known as *rolling hash*. Suppose a window of n characters in size that moves along a document, each time a character, for each position a conventional hash is calculated, which is recalculated very quickly to the next position of the window until at the end of the document. *Rolling hashing* is applied in the search for substrings, among other things. The *fuzzy hasing* applies a *rolling hash* which advancing along a document until it reachs certain points, known as *trigger points*. When it reaches a *trigger point*, the *rolling hash* computed until then is stored and restarts, rolling to the next point, and so on until the end. Both the size *n* of the hash rolling window as the *trigger points* are obtained in various ways depending on the size of the document.

At the end of the process we have a number of hashes corresponding to as many segments of the original document, and we may get a short signature from each hash and concatenate them into a sequence of characters [17]. If we change the original document, one or more of the segments (corresponding to the amendment) will produce different *rolling hashes*, but not the rest. Thus, the final sequence of characters obtained will be different in the corresponding segments affected by the amendment, but not the rest. Thus, the number of different characters in the final *fuzzy hash* is proportional to the magnitude of the changes in the document [10].

The *fuzzy hashing* is used in tasks related to computer security (forensics, detection of polymorphic viruses, malware detection, etc.) [9].

4 Tests Carried Out

We conducted several tests designed to evaluate the usefulness of *fuzzy hashing* to detect duplicate documents on the web. The context of these tests is the Information Retrieval on the Web, in this context, a crawler traverses the network by collecting and indexing documents. Basically, the crawler scans a web page, indexes and extracts the links it contains, which are stored. Once indexed a page, the crawler gets from the storage a new URL to explore. Naturally, the crawler is careful not to scan the same URL, ie, does not store addresses that have been visited or already stored previously. Of course, we can also to limit the tours of the crawler to URLs belonging to certains domains or having certain characteristics, as is also possible to confine the indexing to documents in certain formats, etc. [6].

In the tests we used our own crawler, designed to be part of our search engine VII [7]. We have added the ability to detect exact duplicates, using a MD5 hash, and also to detect similar documents using *fuzzy hash*, using *ssdeep* python module [13], a port for this language based on the *ssdeep* library's Kornblum [11]. In all cases, both *fuzzy hash* and MD5 hash are obtained not from the original page or document, but from its conversion to plain text. In this way we ignore differences arising simply the format. On the other hand, as mentioned before, the crawler only visits different URLs, ie, we refer to duplicate or near duplicate documents with different network addresses.

Since these are preliminary tests, we have carried out crawling inside certain web servers or certain network subdomains. Duplicates found, therefore, are always on the same server or in any case, within the same subdomain. After several tests, we established a similarity threshold of 0.9 for the *fuzzy hash*, so that documents with a score above this threshold are considered duplicates. Some results are shown in the accompanying graphs.

The first thing to note is the abundance of duplicate documents in all cases. Many of them are exact duplicates (conversion to plain text), detected with MD5.

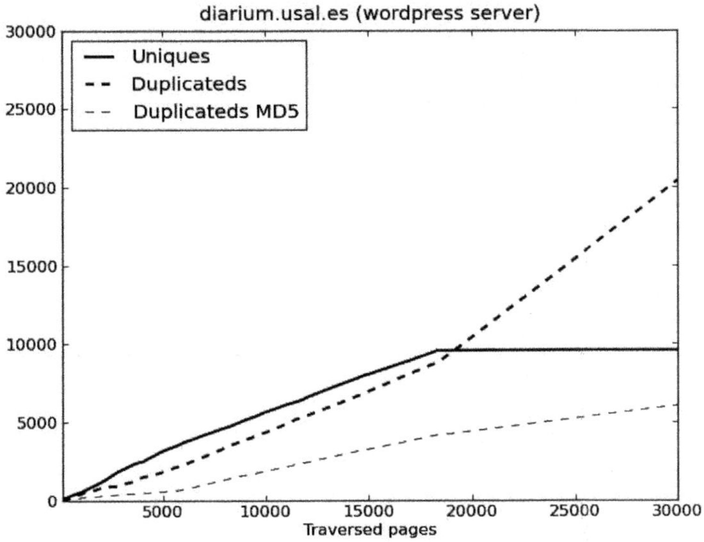

Fig. 1 `diarium.usal.es`. *WordPress* Server

But, in different ways, as the case, there is a significant amount of documents not identified as duplicates by MD5, but thanks to the *fuzzy hashing* can be considered as duplicates for practical purposes. In many cases, the amount of duplicates is even higher than that of unique documents. In some graphs is shown clearly how, as he goes the route of the crawler through a server, the number of unique pages or documents not duplicates grows to a ceiling which is not exceeded or exceeded with little slope. The number of duplicate and near duplicates is also growing as the crawler advances in the exploration of the site, but does not stop and overcomes, sooner or later, to the unique documents.

In other words, the crawler gets all unique documents but does not stop, then continue to get URLs that, being new and different from those already explored, lead to pages or documents that are duplicates of others already visited. This means that quite closely, we may suspend crawler travel long before having explored all the addresses collected, confident that after a certain point, little or no new content page will be covered.

On the other hand, tests have been made also with a full subdomain, the subdomain is heterogeneous in the sense that it consists of multiple servers (about 200) of various sizes, some with static and other with dynamic pages, managed by a variety of *CMS*s and / or ad-hoc Web applications. The size, number of pages, in that subdomain is difficult to determine, Google gives a figure of 1,500,000 pages, although it is known that the figures provided by the search engines must be observed with caution [1]. What is striking is that, after exploring the first nearly 73.000 pages the

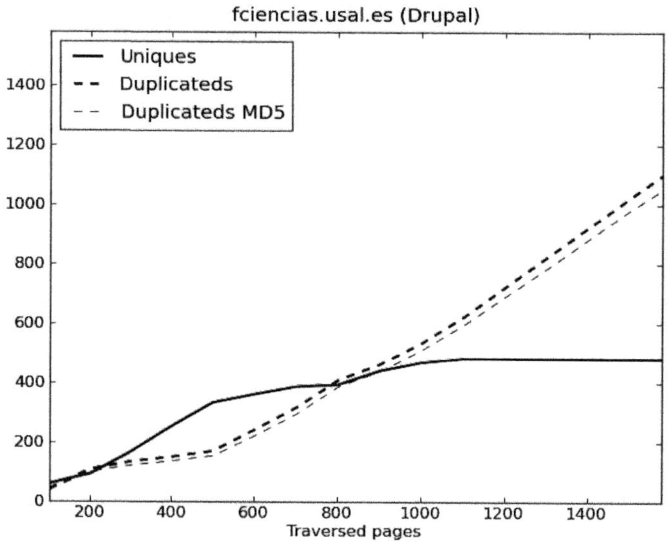

Fig. 2 `fciencias.usal.es`. *Drupal*

crawler don't find unique documents and, of these, only about 46,000 are uniques. The amount of exact duplicate documents (detected by MD5) is remarkable, but also stabilizes after a certain point. Which increases from that point the amount of duplicate documents are near duplicates detected by *fuzzy hash*.

At this point several questions may arise. First, the causes of the extensive duplication of documents. Taking into account that the crawler only follows unique URLs, some of these duplications could come from redirects and aliases in the server names. But, waiting to confirm this hypothesis with hard data, it appears that this cause could only explain part of exact duplication. Another part may stem from how CMSs and web applications generate links to their components, which could take different forms in regard to the URL, but point to the same document. In this regard it is interesting to ask if there are different behaviors depending on the particular software used in each site. We can assume that, perhaps, a CMS generates more duplicates than others. It is also likely that certain web applications generate a greater number of replications, either exact or either generating versions of the same document. This may occur with social-oriented web sites (web 2.0) and is also likely that this class of sites have a greater relationship with certains CMSs, aimed precisely at the Web 2.0.

The way the crawler traverses the site has to do with finding more or less duplicates? Can we find paths to allow us explore before the unique documents contained in a web site or domain?

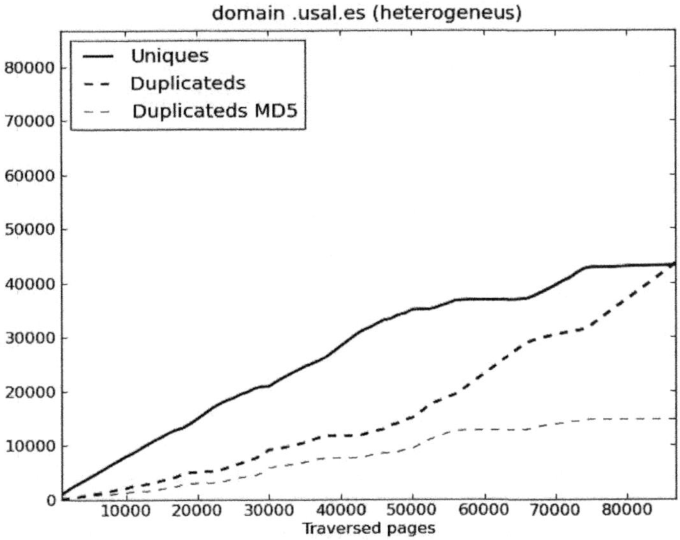

Fig. 3 Domain `usal.es`. Several servers and heterogeneus web applications

5 Conclussions and Future Work

We have shown a technique known as *fuzzy hash* and its use in detecting near duplicates in the web when a crawler navigates it. *Fuzzy hashing* is a real hash, it is computed quickly and the risk of collisions (two different documents produce the same hash) is very low. However, the difference between two *fuzzy hashes* is proportional to the difference between the documents from they came. This feature allow us to detect near dupplicated documents easily. Preliminary tests carried out show that there is a significative amount of duplicates or near duplicates documents. The results suggest that, for a given level of crossed pages within a website, there comes a point where they no longer reach new pages, which have not been explored previously.

For future work, various questions must be resolved. First, it is necessary to analyze the causes of the large number of duplicate documents. Also to consider whether certain CMS produce more duplicate than others. Also, find ways of improving crawling strategies, so that the useful pages can be explored before.

References

1. Bar-Ilan, J.: Expectations versus reality – search engine features needed for web research at mid 2005. Cybermetrics 9(1) (2005), `http://www.cindoc.csic.es/cybermetrics/articles/v9i1p2.html`

2. Bharat, K., Broder, A.: Mirror, mirror on the web: A study of host pairs with replicated content. Computer Networks 31(11-16), 1579–1590 (1999), http://citeseerx.ist.psu.edu/viewdoc/download?doi=10.1.1.90.1488&rep=rep1&type=pdf
3. Chowdhury, A.: Duplicate data detection (2004), retrieved from http://ir.iit.edu/~abdur/Research/Duplicate.html, http://gogamza.mireene.co.kr/wp-content/uploads/1/XbsrPeUgh6.pdf
4. Chowdhury, A., Frieder, O., Grossman, D., McCabe, M.: Collection statistics for fast duplicate document detection. ACM Transactions on Information Systems (TOIS) 20(2), 171–191 (2002), http://citeseerx.ist.psu.edu/viewdoc/download?doi=10.1.1.5.3673&rep=rep1&type=pdf
5. Damerau, F.: A technique for computer detection and correction of spelling errors. Communications of the ACM 7(3), 171–176 (1964), http://www.cis.uni-muenchen.de/ heller/SuchMasch/apcadg/literatur/data/damerau_distance.pdf
6. Figuerola, C.G., Alonso Berrocal, J.L., Zazo Rodríguez, Á.F.: Rodríguez Vázquez de Aldana, E.: Diseño de spiders. Tech. Rep. DPTOIA-IT-2006-002 (2006)
7. Figuerola, C.G., Gómez Díaz, R., Alonso Berrocal, J.L., Zazo Rodríguez, A.F.: Proyecto 7: un motor de recuperación web colaborativo. Scire. Representación y Organización del Conocimiento 16, 53–60 (2010)
8. Hamming, R.: Error detecting and error correcting codes. Bell System Technical Journal 29(2), 147–160 (1950), http://www.lee.eng.uerj.br/~gil/redesII/hamming.pdf
9. Kornblum, J.: Identifying almost identical files using context triggered piecewise hashing. digital investigation 3, 91–97 (2006), https://www.dfrws.org/2006/proceedings/12-Kornblum.pdf
10. Kornblum, J.: Beyond fuzzy hash. In: US Digital Forensic and Incident Response Summit (2010), http://computer-forensics.sans.org/community/summits/2010/files/19-beyond-fuzzy-hashing-kornblum.pdf
11. Kornblum, J.: Fuzzy hashing and sseep (2010), http://ssdeep.sourceforge.net/
12. Levenshtein, V.: Binary codes capable of correcting deletions, insertions, and reversals. Soviet Physics Doklady 10(8), 707–710 (1966)
13. Milenko, D.: ssdeep 2.5. python wrapper for ssdeep library (2010), http://pypi.python.org/pypi/ssdeep
14. Navarro, G.: A guided tour to approximate string matching. ACM computing surveys (CSUR) 33(1), 31–88 (2001), http://citeseerx.ist.psu.edu/viewdoc/download?doi=10.1.1.96.7225&rep=rep1&type=pdf
15. Soukoreff, R., MacKenzie, I.: Measuring errors in text entry tasks: an application of the levenshtein string distance statistic. In: CHI 2001 Extended Abstracts on Human Factors in Computing Systems, pp. 319–320. ACM, New York (2001), http://citeseerx.ist.psu.edu/viewdoc/download?doi=10.1.1.22.757&rep=rep1&type=pdf

16. Tan, P., Steinbach, M., Kumar, V., et al.: Introduction to data mining. Pearson Addison Wesley, Boston (2006), http://www.pphust.cn/uploadfiles/200912/20091204204805761.pdf
17. Tridgell, A.: Spamsum overview and code (2002), http://samba.org/ftp/unpacked/junkcode/spamsum
18. Tridgell, A., Mackerras, P.: The rsync algorithm (2004), http://dspace-prod1.anu.edu.au/bitstream/1885/40765/2/TR-CS-96-05.pdf
19. Yerra, R., Ng, Y.: Detecting similar html documents using a fuzzy set information retrieval approach. In: 2005 IEEE International Conference on Granular Computing, vol. 2, pp. 693–699. IEEE, Los Alamitos (2005), http://faculty.cs.byu.edu/~dennis/papers/ieee-grc.ps

An Implementation of a Trading Service for Building Open and Interoperable DT Component Applications

Carmelo Maturana, Antonio Jesús Fernández-García, and Luis Iribarne

Abstract. Integration and interoperability of software components and information systems is extremely important nowadays to manage data and to integrate applications and information, and to improve business processes in companies. Recently is emerging a new market of software component for Digital Television industry. These components require the development of new methodologies for their integration and interoperability with current and traditional components and web-based information systems (WIS) [2][4]. This paper presents an implementation of a trading service for DT software components as part of a methodology for integrating business applications. The paper also discusses some use-case experiments to validate the proposed implementation.

Keywords: DT software components, trading, interoperability.

1 Introduction

Cloud Applications are awakening these days. These applications are now replacing desktop applications that you used to have to install, configure and maintain. With only a browser, a user can access to completely functional web applications. In order to work properly and prevent communications problems, it is necessary that the service follow the rules of the standards. XML-Schema is the standard of W3C to integrate and share information on Internet. We use XML-Schema in this project. The world implementation of Digital Television not only allows optimal and more digital channels. Digital Television has emerged technologies that will shortly produce new markets of software components. At the present time, some companies such as Boxee box, Apple TV, Roku and now Google TV [5] are working in this field. Their solutions are based on connecting the Television to the Internet via a

Carmelo Maturana · Antonio Jesús Fernández-García · Luis Iribarne
Applied Computing Group
University of Almeria, Ctra. Sacramento s/n
e-mail: `luis.iribarne@ual.es`

J.M. Corchado et al. (Eds.): Trends in PAAMS, AISC 90, pp. 127–135.
springerlink.com

proprietary Operating System. For instance, Google TV applications do not run in any Set-Top-Box, only in Set-Top-Box with the specifics APIs that Google provides to developers. For that reason, they can take advantage of their OS Android applications but they have the disadvantage that these applications can not be used in other systems and most people cannot use them. In this paper, we describe an open source implementation based on Interactive Digital Television standards (DVB project [14]) that work in any standard Set-Top-Box.

The component-based software engineering partly solves the "interoperability" by means of mechanisms of trading (traders) for COTS components (commercial components) [6]. A trader can be defined as a trading service between software components. It allows the cooperation of different components to the sequentially data processing in order to obtain information or create a final service more complex. These components are in a repository where information is stored and maintained [7]. To know how the component market works we have studied similar commercial sites we can find in Internet at this moment. We analyzed 13 large stores according to 20 criteria. Figure 1 shows a short view of 4 large stores according to 8 criteria.

This paper is organized as follows. Section 2 presents the project description. Section 3 shows the project implementation. Finally, Section 4 presents some conclusions and future work.

WEB / CRITERIA	ANDROID MARKET (GOOGLE)	GOOGLE TV (GOOGLE)	APPSTORE (APPLE)	APPLE TV (APPLE)
Access for different platforms	(V)	X TV + keyboard and special receiver	(V)	X TV + Apple TV device
App for various OS	X Only Android (mobile)	(V) -Tv content -Android	X Only Mac (mobile - laptop - desktop)	(V) -Tv content -iPhone OS
Each access platform optimization	(V) Web - Android - DTT	-	(V) Web - iPhone	-
Committed store depending on the platform	**Web** -Publication -Additional information **Android** -Access to developers -Purchase/download/use -Ratings, opinions **TDT** -*(GOOGLE TV)*	Everything on TV	**Web** -Publication and advertising -Apple views **iPhone** -Purchase / download / use -Ratings, opinions	Everything on TV
Content development tools	(V) Free IDE	(V) IDE - owners and free	X IDE private (only are permitted app with IDE-Apple)	X IDE private (only are permitted app with IDE-Apple)
Exclusive distribution	X Applications are not necessarily linked to this store	No data currently	(V) Any application must be certified by Apple to run on the mobile device	(V) Any application must be certified by Apple for their implementation
Restrictive policy for the admission of app	Mid-level (V) Fast repository growth X Very variable in app quality	(Android and Google TV share app)	High level (V) Medium high-quality X Slow repository growth	(Apple TV and iPhone share app)
Available from television (DTT)	(V) *(Forthcoming with GOOGLE TV)* *	(V)	X	(V)

Fig. 1 4 large component stores description

2 Project Description

In this paper we present a personal implementation of a trader according to our necessities [3]. In order to understand how it works it is necessary to know what a service type is. A Service Type (ST) is an entity, with properties, that represent similar functionalities. A ST can inherit from other or others services types. Inheritance is one of the most powerful characteristics of a trader, which allows component cooperation. Different STs can work together in order to give a wider service. We offer applications for DT. In our trader a component is a service that we offer to the clients. All offers inherit from a ST their properties and functionality. For example, a component A that adds real numbers, could inherit from a ST called "real adder" and other component B that subtract real numbers could inherit from "real subtracter". Both ST could be joined obtaining as a result other ST called "real calculator"; this service inherits from "real adder" and "real subtracter". The granularity of ST is defined by the users that are creating them.

Other powerful characteristic of a trader is the ST hierarchy. When registering a ST it is possible to specify another ST that, semantically, has the same functionality or wider. For instance, a developer creates a component to "add" and "subtract" real numbers. He/she registers the service in the trader with the name "integer calculator" and he/she could indicate to the ST that "real calculator" is higher in the same hierarchy. Thus, when a client asks to the trader for an "integer calculator" service the trader can offer the "integer calculator" and alternatives services like "real calculator" with possibilities that could be useful to the client. Besides, the trader checks the interoperability between two components so the output of one of them can be the input of the another one obtaining progressively richer information and wider services. The interoperability information is associated with each offer and can be found in the *Interface Repository*. The standard of a trader has three main independent repositories: RO, IR and STR. Our trader object uses four independent repositories (we added a DataRO repository):

- RO (*Repository Offers*). In this repository developers make the register of their components as instances of a ST.
- IR (*Interface Repository*). In this repository are stored the interfaces of the offered components. A service exported by a client has a computational interface which responds to the functionality of the service.
- STR (*Service Type Repository*). Where the STs are Stored.
- DataRO (*Data Repository Offers*). It has additional information associated to services (offers), i.e., the source code (that can be stored in this repository or in it can be a link to a external repository), use license, etc.

Figure 2 shows the Systems Architecture and the repositories structure, technologies, tools and standard used. The trader is hosted in a XML Web Server. There is a virtual proxy in each side (client and server) that automates the encapsulation/decapsulation of the messages according to the data and metadata structures of a SOAP message [15] creating a XML file [13]. The client and server communication flow as follow:

- **Client Object.** User Interface to access to the Register and Lookup interfaces.
- **Client Proxy (asp.net).** The "offer register" and "look up" services SOAP messages are encapsulated. The list of services registered and the result of a search services query are decapsulated.
- **Internet**. SOAP messages.
- **Server Proxy (C Sharp).** The offer register and look up services SOAP messages are decapsulated. The list of services registered and the result or a search service query are encapsulated.

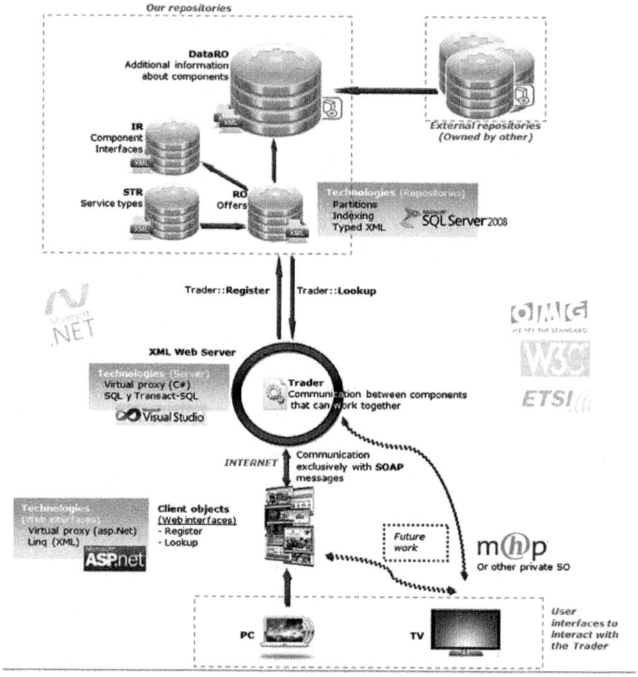

Fig. 2 System Architecture with standards and technologies

- **XML Web Server.** It handles the trader and the Register and Lookup interfaces.
- **Repository.** The four repositories (RO, IR, STR, DataRO) according to the XML-Schema standard, which can be stored in different servers.

3 Implementation

The software used for the development is: Visual Studio 2010 (C# and ASP.NET), SQL Server 2008 Enterprise (SQL y Transact-SQL). It is required the enterprise version to use advanced organization functions for a better optimization. FrameWork .NET 3.5 and Windows Server 2008 for the server. Let's now see some grammars, repositories, XML Web Server, and Trader interface and policies details.

Grammars: We want this project to be distributed (client-server with access to independent repositories) and multiplatform (there are different OS in Digital Television Applications). Therefore, all the information will be stored in a repository as XML files. Each repository store data about the system such as: trader, COTS and Operating System Platform information. Each component has associated templates where their functionality, features and location are defined as well as their interoperability with other components. A grammar (scheme .xsd files) is written for the ST (STR) as well as other grammar where additional information such as icons, pictures, source code, users opinions, etc., is stored.

Once we have defined the basic structure of storage, the next step is to know the common data of the software components. For that, we have researched how the Eclipse and Mozilla Firefox Pluging works and we have extracted elements such as component size, owner, use license, etc. This information is stored in the RO (*Repository Offer*) and the DataRO (*Data Repository Offer*) repositories. The information used in the "lookup" interface is stored in the RO and the additional information in the DataRO. It is necessary to know the nature of the components that will be managed by the trader. In this case, this trader handle software components to offer services (applications) for the Interactive Digital Television and we have studied MHP[10][8]. MHP (*Multimedia Home Platform*) is the standard specification to develop Interactive Services for Digital Television. In the review of MHP we have found some useful information to find components:

- **BAT** give the possibility of search component for a specific channel, program or from a specific transmitter.
- **Profile and version** are needed in order to know the interoperability between components and to know if a Xlet (a MHP component) is signed or not. The standard only allows communications between signed or not signed applications.
- **Parental Control.** We can not forget that we are talking about TV applications and they have emission laws. The grammar have to admit parental control that allows parents or other caregivers to block programming on their televisions that they do not want children to watch.

Our system is ready to work with standard MHP components but monitoring the market we have found a lot of private systems who has their own OS for Interactive Services in Digital Television. Noticing that, the grammar has been written according to this fact. The grammar is ready to store component information that only make sense in a particular OS and it saved only if the component belong to this OS.

Repositories: To create the repositories we have attended to efficiency and consistency measures.

- **Efficiency.** The partitions allow to divide a table in different files [12]. The records are distributed in files according to the value of a property following the rules of a partition scheme. The XML files are indexed under a primary index. Secondary indexes are optional.
- **Consistence.** Typed XML. A XML instance is associated with a grammar. The DB engine always checks the consistency between the instance and the grammar in insertions and modifications.

The repository has been created in 4 steps:

Step # 1. *Database creation and partition files*: We have 17 categories but we create 18 files (one for future uses) for each repository (excepts STR). In total we have 57 files: 2 DB organization files + 1 log file + 18 RO files + 18 DataRO files + 18 IR files (STR files are not necessary).

```
CREATE DATABASE repository
--File Groups for the Repository. (Offers)
FILEGROUP RO1 ( NAME = RO1dat1, FILENAME = 'C:/REPOSITORY/RO1dat1.ndf'),
---
FILEGROUP RO18 ( NAME = RO18dat1, FILENAME = 'C:/REPOSITORY/RO18dat1.ndf'),
parent.isPlayed.name = self.parent.isPlayed.name) )
```

Step # 2. *Creating scheme collection*:

```
CREATE XML SCHEMA COLLECTION XSD_RO AS N'<GRAMMAR>';
-- Verification. List of the schemes collection of the DB.
select * from sys.xml_schema_collections
-- Verification - List of namespaces in DB.
select name from sys.xml_schema_namespaces
```

Step # 3. *Creation of the partition function and schemes for the repositories*: The function partition split each repository into pieces (group files). The attribute used to create the partition function is "category". There is only one partition function because "category" is used in all the repositories. Each partition uses the partition function with the scheme partition to associate every record in a specific repository.

```
-- Partition Function according to CATEGORY.
CREATE PARTITION FUNCTION PF_category (int) AS RANGE LEFT
FOR VALUES (1,2,3,4,5,6,7,8,9,10,11,12,13,14,15,16,17);

-- Partition Scheme for the offer repository.
CREATE PARTITION SCHEME RO_categoryPS AS PARTITION PF_category TO ( RO_1, --, RO_18 );
```

Step # 4. *Creation of Repositories, Partitions and Indexes*: Figure 3 shows the repositories and their relations. The RO repository keys are "id_service" and "category_range" because the column used in the partition function has to be part of the key. In this project the use of partitions do not difficult the design and implementation but improve the system.

```
template XML (XSD_RO) not null, -- typed XML
-- Partitions. According to the function partition defined before and the scheme partition
ON RO_categoryPS (category_range);

-- Primary XML index. Indexing Offers. Offers are written in XML documents.
CREATE PRIMARY XML INDEX index_XML_RO ON RO(template);
```

XML Web Server: The XML Web Server inherits from the proxy class. *Soap Header* derives from *SoapHeader* class (see next table).

```
public class TemplateHeader : SoapHeader {
public string username;
public string password;
public DateTime created; }
```

The SOAP message creation in the client object consist in the instance of one of the SOAP headers that we have defined in the Server. Then, create a new instance of the Proxy class from the libraries loaded in the Server. The client knows the server and it has access to his libraries through his namespace. A header is added to the

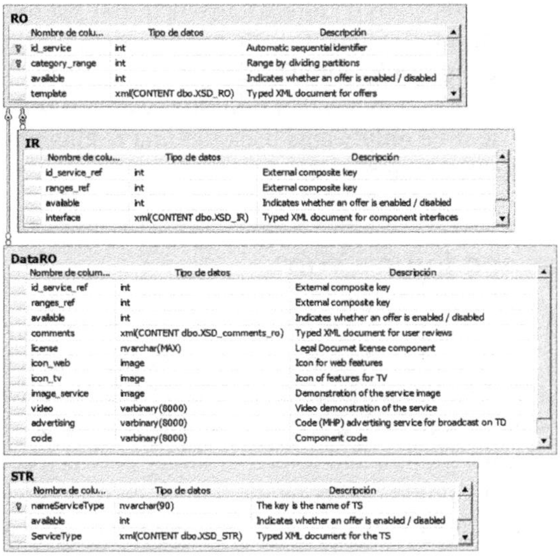

Fig. 3 Repository Relations

Proxy and a method of the web service is called through the Proxy class instantiated. Then, a decapsulated SOAP message is returned (next table).

```
TemplateHeader TemplateHeader_soap = new TemplateHeader();
TemplateHeader_soap.username = "username";

dummy_webservice.WebServiceTrader proxy = new dummy_webservice.WebServiceTrader();

proxy.TemplateHeader_aux = TemplateHeader_soap;
x = proxy.List_ServicesTypes();
```

In the trader a SOAP message is received from the client. The trader access to his data in the repository and send back to the client through his proxy the results obtained. Possible results can be a set of offers, errors in the parser XML types, etc: `SoapHeader("TemplateHeader_aux")] public string MethodWeb(){ }`

Trader Interfaces "Register" and "LookUp": A client can interact with the trader via "register" or "LookUp" user interface. In **LookUp** interfaces the client ask the trader for services.

Trader Policies: The main responsibility of the trader is to satisfy the search of services from the clients. These searches are focused in ST and other 3 topics:

- **Constraints.** Boolean expression written in OCL language. The properties of a ST are inherited for the component that inherits from this ST.
- **Preferences.** Indicates the order in which a client wants the trader offer the services: max (ascending), min (descending), with (condition), random (randomly) and first (in natural order). In the trader definition the order is defined by itself,

however, in our implementation we have decided that the clients choose some preferences (max and min). Thus, the client can interact in real time with the trader to obtain different results for his searches.

- **Searches.** There are some policies that limit the search of services in a query so the client can choose if he only wants back the first results found or he wants to search in only one part of the repository.

4 Conclusions and Future Work

The market of software components is growing and interactive applications for Digital Television too. In this project we propose a trader to look for components and provide them to clients and a marketplace for developers to register their components. We do not want only ask for a component and give back that component, we pretend to exploit the integration and interoperability between components by creating new services using single components that can cooperate and work together in order to give a satisfactory experience to the client. Once created the trader and the interfaces "Register" and "LookUp" the next step is to increase the number of the functional interfaces of the trader and improve them to offer a practice, simple and stylish interface where the clients could access to every aspect of the trader in a easy way. It is also necessary to increase the applications stored in the repository by the development of interactive applications and promoting the trader so developer can register their components.

Acknowledgment. This work was funded by the EU and the Spanish MICINN under grant of the TRACE project TRA2009-0309 and Ingenieros Alborada IDI.

References

1. Amerini, I., Caldelli, R., Becarelli, R., Filippini, F., Ballocca, G., Borri, R.: Integration between digital terrestrial television and Internet by means of a DVB-MHP web browser. In: WEBIST 2009, pp. 323–328 (2009)
2. Claras6, J.A., Baldo, D., Benelli, G., Daino, G.L., Zambon, R.: Interactive Digital Terrestrial Television: The Interoperability Challenge in Brazil. J. of Dig. Mult. Broad, 17 (2009)
3. Fernández-García, A.J., Iribarne, L., 83–88: TDTrader: A methodology for the interoperability of DT-Web Services based on MHPCOTS software components, repositories and trading models. In: IWAAL CEDI 2010, Valencia, pp. 83–88 (2010)
4. Gallego, D.A., Ballano, A.E., Vea-Murguia, J., Navamuel, J.J., de la Hoz, P.M., Navarro, A., Lafuente, A.: Multiplataform environment joining digital TV, mobile devices and traditional e-learning with collaborative learning. In: 1st ICCSE 2009 Lisbon, Portugal, pp. 175–184 (2009)
5. Google TV Project, `http://www.google.com/tv`
6. Iribarne, L., Troya, J.M., Vallecillo, A.: A trading service for COTS components. Computer Journal 4(3), 342–357 (2004)

7. Iribarne, L., Troya, J.M., Vallecillo, A.: Trading for COTS Components to Fulfil Architectural Requirements. In: Development of Component-Based Information Systems: Advances in Management Information Systems, vol. 2, pp. 202–222. M.E. Sharpe (2005)
8. KDB, The MHP Knowledge Project (KDB), http://www.mhp-knowledgebase.org
9. Melendreras-Ruiz, R.: ImplanTDT: Usability laboratory, real user DTT monitoring platform and MHP-based services. In: 5th IEEE CCN, pp. 249–250 (2008)
10. MHP 1.2 Specification (2009), http://www.mhp.org
11. MSDN, Indexes on XML Data Type Columns, http://msdn.microsoft.com/
12. MSDN, Partitioned Table and Index Concepts, http://msdn.microsoft.com/
13. MSDN, Managing XML Schema Collections on the Server, http://msdn.microsoft.com/
14. OMG, The OMG Trading Object Service, http://www.omg.org/cgi-bin/doc?formal/00-06-27.pdf
15. SOAP Version 1.2, W3C, http://www.w3.org/standards/techs/soap

Towards Semi-automatic Construction of Data Mashups

Javier López, Fernando Bellas, Alberto Pan, and Paula Montoto

Abstract. Mashup applications combine pieces of functionality from several existing web data sources to provide new integrated functionality. In this work-in-progress paper, we outline a new method for creating data mashups that suggests to users new relevant data sources and the most suitable ways of combining them. This makes the data mashup process much easier for non-technical users.

Keywords: Enterprise Mashups, Integration Patterns, Web Engineering.

1 Introduction

Mashup applications combine pieces of functionality from several existing websites and web services to provide new integrated functionality. Some key differences between mashups and conventional integration approaches are:

- "Power users" without programming skills should be able to create mashups to address tactical needs.
- Strong focus on collaboration. Users create mashups by reusing components and mashups created by other users.

The problem of creating mashups can be divided into two main subproblems: 1) Create 'data mashups' that combine existing data sources and return unified data, and 2) Create the final graphical interface for the mashup by composing 'widgets' that can be populated by the data mashups.

Most existing methods [4] for creating data mashups provide a graphical interface to easily combine data from several web services. In this work-in-progress

Javier López · Fernando Bellas · Alberto Pan · Paula Montoto
Information and Communications Technology Department, University of A Coruña
Facultad de Informática, Campus de Elviña, s/n, 15071, A Coruña, Spain
e-mail: {jmato,fbellas,apan,pmontoto}@udc.es

J.M. Corchado et al. (Eds.): Trends in PAAMS, AISC 90, pp. 137–144.
springerlink.com

paper, we outline a new method for creating data mashups that goes one step beyond: at each step of the creation process, the tool suggests new relevant data sources and the most suitable ways of using them in the current mashup.

Section 2 of the paper explains the idea with an example. Section 3 overviews our approach. Sections 4 and 5 respectively describe its two main components: how to suggest relevant sources and how to determine the best way to use them in the current mashup. Section 6 concludes the paper and outlines our future work.

2 Running Example

Martin works as Sales Manager at Acme Inc. Acme uses the on-demand CRM Salesforce (http://www.salesforce.com) to store data about its clients. Customer data includes a 'Satisfaction Level' field. Martin has an idea to discover new potential customers: querying Salesforce for satisfied clients and using LinkedIn (http://www.linkedin.com) to find their business contacts. The idea is that Martin can call them and use the satisfied customer as reference.

Therefore, Martin starts the data mashup creation tool and drags and drop on the workspace the 'Salesforce' and 'LinkedIn' components. These components allow executing queries on the web services provided by such companies. The 'Salesforce' component has been provided by the IT department of the company while the LinkedIn component has been contributed by another user with some programming skills. Martin combines the two components using the 'Enrich' operation, which is conceptually similar to a relational join, and renames the resulting component as 'SatCustContacts'.

Then, the tool automatically makes several suggestions to Martin. He finds useful two of them: the first one is 'Merge LinkedIn with Xing': the 'merge' operation is conceptually similar to the 'extended union' relational operation and the tool suggests mappings between fields that represent the same information in both sources (Fig 1). Therefore, this will allow Martin to look for satisfied customers' contacts in another prominent business network.

Martin also takes another of the tool suggestions: 'Enrich SatCustContacts with YahooFinance'. This will allow Martin to enrich the new potential clients' data with additional information about the companies they work for (in this case, the

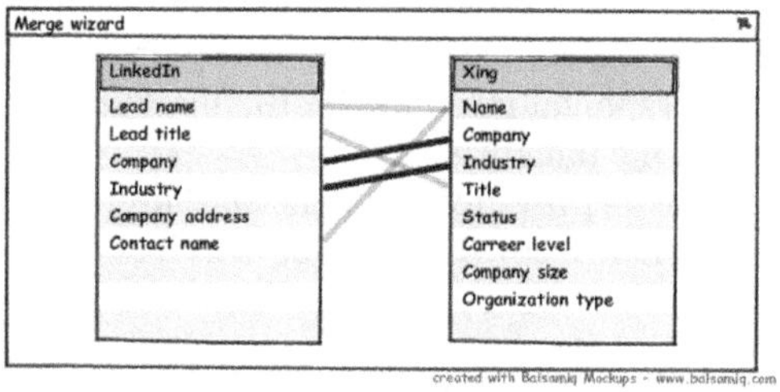

Fig. 1 Mappings between 'LinkedIn' and 'Xing'

tool suggests mapping the fields representing company information in both sources). The resulting mashup is shown in Fig 2. At any time, Martin can preview the results of the current mashup by using the 'Test' area.

After these steps, the tool keeps doing new suggestions: now it is suggested to 'Merge YahooFinance with GoogleFinance' to obtain even more information about contacts' companies. But Martin is already happy with the mashup results, and finishes the process.

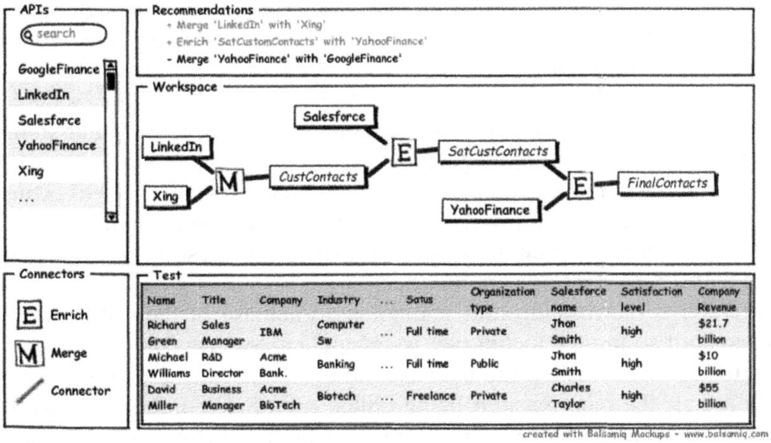

Fig. 2 Resulting mashup

3 Overview of the Recommendation Architecture

The suggestions in our example are obtained through a two-fold process: 1) Determining new relevant data sources to be included in the mashup, and 2) Determining how to combine the data source with the current mashup.

The first task is done by applying collaborative filtering techniques [1] to find data sources frequently used in previous mashups together with the data sources in the current mashup. Section 4 provides the details.

The second task receives as input two sets of data sources: those produced by the first task and those in the current mashup. For each pair of one recommended data source and one data source in the mashup, it performs the following tasks: 1) Determine what data combination operation(s), if any, can be applied to combine both data sources, 2) Determine how the data operation should be configured (e.g. determining the 'join' fields in the 'enrich' operation), and 3) Estimating a confidence score for the recommendation.

To perform this second task we use 'schema matching' techniques to find semantic mappings between fields in both data sources. For instance, if we are able to find semantic mappings between almost all fields of the two data sources, then they probably return the same kind of data and it makes sense to apply the 'merge' operation between them. In turn, if we find semantic mappings for only one or two data fields, then the two data sources probably return different kinds of data that can be joined by the mapped fields. Section 5 describes the process in detail.

4 Suggesting Relevant Sources

4.1 The General Approach

To recommend relevant sources to the user we have studied two techniques based on *Collaborative Filtering* (CF) [1]. CF has been extensively used for product recommendation in e-commerce. In the e-commerce context, CF-based recommenders maintain a user-item matrix of scores (from 0 to 5, for example) with many missing values. Each score represents the interest of a user in an item (product). When the system recommends products to the user, it suggests her x products not rated previously by her. To do so, CF-based recommenders try to estimate the ratings of the products with missing values in the row corresponding to the user u taking into account the ratings of the full matrix. Once estimated, the x products with higher estimated ratings are returned.

Intuitively, if the mashup tool maintains a database of the data mashups that have been built so far, CF techniques seem applicable for recommending new data sources for the data mashup the user is creating. Consider the example introduced in section 2. If the data mashup being edited uses Salesforce and LinkedIn, and there are a significant number of data mashups built in the past that use Salesforce and Xing or LinkedIn and Xing, it seems reasonable to suggest Xing. To make this possible, the tool automatically maintains a mashup-source matrix as users create mashups. Each row represents the sources a data mashup has used. Whenever the user requests the tool to recommend her new sources, the tool estimates ratings for the rest of sources not currently used by the current mashup taking into account the mashup-source matrix, and return the *top-x* ones with higher ratings.

More formally, let $M = \{m_i\}_{i=1..n}$ the set of mashup that were built with the mashup tool, $S = \{s_j\}_{j=1..m}$ the set of possible data sources, and $R = \{r_{ij}\}_{i=1..n,\, j=1..m}$ the mashup-source matrix. Each value r_{ij} in R represents if m_i has used s_j. Let m be the mashup being edited and $S(m) \subseteq S$ the set of sources that it currently uses. Taking into account the matrix M, the recommender must estimate the ratings r_{ms} of the sources $\{s_j \;/\; s_j \in S \text{ and } s_j \notin S(m)\}_{j=1..m}$ for the mashup m, and return the *top-x* sources with the higher estimated ratings.

We adapted two famous CF algorithms for estimating the ratings: k Nearest Neighbors [1] and Naive Bayes [5]. Details on how the algorithms were adapted are not provided due to space limitations and will be provided in a future work.

4.2 Evaluation

To evaluate the recommenders we have implemented, we selected a high-quality subset of the sources and mashups published in ProgrammableWeb (*http://www.programmableweb.com*). ProgrammableWeb is one of the main mashup public directories. On March 2010, there were around 4600 mashups registered and around 2000 sources (called "APIs"). After discarding collage-like mashups and sources not providing a data interface, we concentrated on the most popular (used) sources and the mashups using three or more of these sources exclusively (we did not remove any source from any mashup, since this would affect the mashup's semantics). This resulted in 29 sources and 157 mashups selected. On average, each mashup used 3.6 sources and there were 6.7 sources that could be correctly recommended for each mashup. There were less than 3 recommendable sources for the 23% of mashups. There were 2 mashups with no recommendable sources.

To evaluate the recommendation performance of the implemented recommenders, we evaluated by hand all sources for all the mashups and constructed a list for each mashup containing the recommendable sources. To run the evaluation, we implemented a leave-one-out [5] evaluation algorithm. In each iteration, a row (mashup) of the matrix was considered as the test set and the rest of the matrix as the training set. For each mashup, we requested the recommender to recommend x_i (top-x_i) sources, being x_i the minimum between 3 and the number of recommendable sources for the mashup m_i (bear in mind that 23% of mashups had less than 3 recommendable sources). After running all the iterations, we calculated three quality measures, *P* (precision), *R* (recall), and *U* (utility), defined as:

- $P = tp \,/\, (tp+fp)$, being *tp* (true positives) the number of sources correctly recommended and *fp* (false positives) the number of non-recommendable sources emitted.
- $R = tp \,/\, \Sigma_{i=1..n} xi$.
- *U = (number of times the recommender emitted at least one recommendable source among the emitted ones for each mashup) / (number of mashups with one recommendable source at least).*

To analyze the quality results of each recommender we considered a random recommender (*random*) as base case. For the mashup m_i, this recommender emits randomly x_i sources not used by m_i.

kNN recommenders can be tuned by adjusting the number of neighbors (*k*), the minimum number of neighbors, and the minimum similarity required to predict a rating. For this reason, as part of the evaluation we run the leave-one-out algorithm several times for kNN recommenders, varying *k* from 1 to 10, the minimum number of neighbors from 1 to *k*, and the minimum similarity from 0 to 1 in increments of 0.1. We choose the configuration that achieved the best precision. Since it is not possible to maximize precision and recall at the same time, we decided to analyze two optimal configurations for kNN recommenders, one forcing the recall to be greater than 0.49 and another one with a recall greater than the recall of the random recommender (R_r). Potentially, the second configuration may produce better precision results, since R_r is less than 0.49.

Fig. 3 illustrates the precision, recall, and utility achieved by each recommender, as well as the measures for the base case (*random*). For kNN recommenders, there are two columns for precision, recall, and utility. The first one corresponds to a configuration with a recall greater than 0.49 and the second one to a configuration with a recall greater that R_r. As figure 1 shows, kNN recommenders achieve better results than *naïve-bayes*. In particular, one of kNN recommenders (*knn-source-wa*) achieves P=0.65, R=0.65 and U=0.86. The performance of the base case is much lower (P=0.29, R=0.29 and U=0.54). *naïve-bayes* is between kNN recommenders and the base case, achieving P=0.43, R=0.43 and U=0.68.

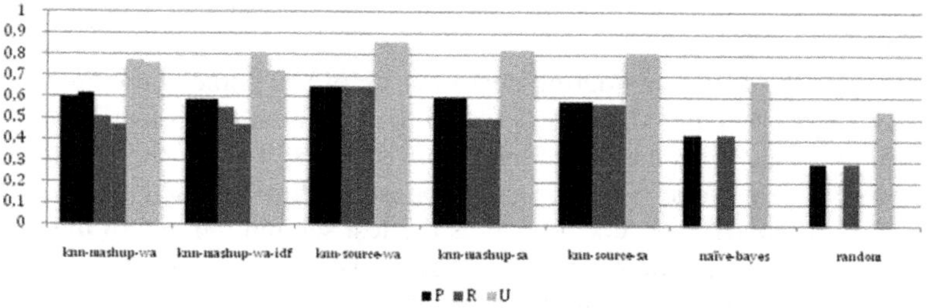

Fig. 3 Precision, recall, and utility for the evaluated recommenders

In conclusion, evaluation results are promising, showing that source recommendation is viable. From the algorithms studied, kNN recommenders seem to perform better than *naïve-bayes*. A data set with a higher proportion of mashups against sources would probably produce better results.

5 Suggesting Source Composition

We consider two main data combination operations:

1. Enrich. Extends the data about an entity in a data source with information from another data source. It is similar to a relational join operation. To execute it, the 'join' fields in both data sources must be mapped. In our running example, the enrich operation was used to add information from contact's companies using Yahoo Finance.
2. Merge. Fuses the result sets obtained from two different data sources to obtain a single result set containing one entity for each entity in the original sources (except duplicates). It is similar to the relational union operation but does not force the two input relations to have exactly the same schema (missing fields are filled in with NULL values). To execute it, the fields in both data sources with the same semantic must be mapped. In our example, this operation was used to merge the contacts obtained from Xing and LinkedIn.

Our tool also allows users to perform other data manipulation operations such as projections (dropping fields, renaming fields and/or applying expressions to fields) and selections (filtering entities by a certain condition), but we think power users can easily perform these tasks by themselves and, therefore, the tool does not make suggestions about them.

The process to determine the best composition operation between two data sources consists of the following steps:

1. Periodically, all the data sources in the system are clustered according to their "schema" similarity. We use HAC clustering [5]. Two schemas are "similar" if they have many fields with the same semantic. To apply clustering, we need a way to estimate the similarity between two schemas. This will be addressed below.
2. If two data sources are in the same cluster, then we consider that the "merge" operation can be applied to them. To apply the "merge" operation we also need to map the fields from both sources with the same semantic. To do this, we estimate the 'field similarity' between two fields (details will be described below). Fields with a similarity greater than a certain threshold will be considered as valid mappings.
3. If the two data sources are not in the same cluster, then we try to find semantic mappings between the fields in both data sources using again "field similarity". If we find them, then we consider that the data sources can be combined with the 'enrich' operation using the mapped fields as join fields. If we do not find any mappings, we do not provide any suggestion.

The following sub-section describes how we estimate "field-to-field" similarity. Once field-to-field similarities have been estimated, "schema-to-schema" similarity between two data sources is computed through a relatively straightforward process that selects the best mapping for each data field in one of the data sources, and then computes the average of those field-to-field similarities.

5.1 Field-to-Field Similarity

We use three measures to compute field-to-field similarity. The final estimate is obtained by combining the evidence obtained from them. The measures are:

1. Textual similarity between the field names. We use the JaroWinklerTFIDF [3] algorithm to estimate the similarity.
2. Similarity between values of the fields. This measure is for character string attributes only. For each data source, the system maintains a set of example tuples obtained from queries executed from previous mashups. Generally speaking, this measure will consider two data fields as similar if their example values share many words in common. We use the standard TFIDF technique.
3. Structural Similarity between example values of the fields. We use a modified version of Dataprog [2] to induce patterns describing the example values of both data fields. For instance, given examples like the following:

   ```
   1 Mi, 1.6 km
   3 Mi, 4.8 km
   ```

   Our modified version of Dataprog could output patterns like:

   ```
   TTN|SMALL "Mi," TTN|DECIMAL "km"
   TTN|NUMBER "Mi" TTN|PUNCT TTN|NUMBER "km"
   ```

The first pattern specifies a small number *(TTN/SMALL)* followed by the constant string *"Mi,"* then a decimal number *(TTN/DECIMAL)* and then the *"km"* constant string. The second pattern is similar but uses the more general tokens *TTN/NUMBER* that matches with any number and *TTN/PUNCT* that matches with any punctuation symbol. Given these patterns, we can estimate the similarity of two fields by matching the examples of one field against the patterns generated from the examples of the other. The estimated similarity will depend on the number of matches and on the restrictiveness of the patterns (matching with a more restrictive pattern scores higher).

To combine the three measures in a single estimation, we first normalize them to make them comparable (details are not described due to space constraints). Once the three values are normalized, we have experimented with a number of functions to obtain the final estimation such as choosing the maximum value, computing a weighted average or using probabilistic approaches. Further experimentation will be done to determine the best option.

6 Conclusions and Future Work

In this work-in-progress paper, we have outlined a new method for creating data mashups that suggests to users new relevant data sources and suitable ways of combining them. This makes the creation of data mashups much easier for non-technical users. We have already implemented and evaluated the algorithms for recommending new data sources, obtaining promising results. Our current work is focused on testing the proposed algorithms for suggesting suitable compositions.

References

1. Bell, R.M., Koren, Y.: Improved Neighborhood-based Collaborative Filtering. In: KDDCup 2007 Workshop, Organized as Part of the 13th ACM SIGKDD International Conference on Knowledge Discovery and Data Mining, pp. 7–13 (2007)
2. Lerman, K., Minton, S., Knoblock, C.: Wrapper Maintenance: A Machine Learning Approach. J. Artif. Intell. Res (JAIR) 18, 149–181 (2003)
3. Cohen, W., Ravikumar, P., et al.: A Comparison of String Distance Metrics for Name-Matching Tasks. In: Workshop on Information Integration on the Web, pp. 73–78 (2003)
4. Yahoo Pipes, http://pipes.yahoo.com
5. Witten, I.H., Frank, E.: Data Mining: Practical Machine Learning Tools and Techniques, 2nd edn. Morgan Kaufmann, San Francisco (2005) ISBN: 978-0-12-088407-0

Adding Semantics to the Discovery of Web Services in Life Sciences

María Pérez, Ismael Sanz, Rafael Berlanga, and María José Aramburu

Abstract. Research in the Life Sciences largely depends on the integration of large, distributed and heterogeneous data sources and web services. Due to the large number of available web services, the sheer complexity of the data and the frequent lack of documentation, discovering the most appropriate web service for a given task is a challenge for the user. In this paper we propose a semi-automatic approach, based on semantic techniques, to assist researchers in the discovery and selection of the most appropriate web services to fulfill a set of given requirements. We describe the overall framework of our approach, and the usefulness of our techniques is demonstrated through a Bioinformatics case study.

Keywords: Web services discovery, users requirements, semantics, Life Sciences.

1 Introduction

Contemporary research in the Life Sciences depends on the sophisticated integration of large amounts of data obtained by in-house experiments (for instance, DNA sequencing) with reference databases available on the web [4]. This is followed by analysis workflows that rely on highly specific algorithms, often available as web services [3]. The amount of data produced and consumed by this process is prodigious; however, the sheer amount of available resources is a source of severe difficulties.

The landscape of Life Sciences-oriented web services is large and complex: there are thousands of available resources [6], but unfortunately only a few are described by adequate metadata. In most cases, their documentation does not provide information about the functionality, the parameters of the

María Pérez · Ismael Sanz · Rafael Berlanga · María José Aramburu
Universitat Jaume I, Spain
e-mail: {mcatalan,isanz,berlanga,aramburu}@uji.es

J.M. Corchado et al. (Eds.): Trends in PAAMS, AISC 90, pp. 145–152.
springerlink.com

service, or even the format of the information it can handle. In break addition, there are many versions of different services that apparently provide the same broad functionality (a particular insidious example is the variety of services providing variants of alignments of genes and proteins), but not enough metainformation is available to decide which of these services is actually the most appropriate for a precise task. There are approaches that have addressed the problem of web service discovery by curating themselves their set of services. For instance, Feta [5], a web service discovery system used by Taverna, bases the discovery of web services on the annotations that have been made manually by a curator using their own ontology. However, it does not provide any ranking to help the user for an appropriate selection of web services. Addressing this problem, [1] tries to discover the best web service based on the intrinsic quality of the service (response time, availability etc.) but does not take into account how appropriate the service is for the actual requirements of the user.

Therefore, the discovery of the web services must be driven by the user requirements and, then, an assessment process is required to determine the most appropriate services. In addition, the user should be involved in the assessment process, and as an expert she should be able to modify its parameters. Moreover, since usually some services provide similar functionality, the assessment techniques must not only evaluate the quality of web services, but also must determine their similarity and that with requirements.

The main contribution of this paper is a semi-automatic approach to assist the user in web service discovery, looking for services that are appropriate to fulfill information requirements in the Life Sciences domain. The mapping between users' requirements and the available web services is based on a semantic normalization of both the requirements and the web services metadata.

The remainder of this paper is structured as follows. Section 2 presents an overview of the proposed approach. In Section 3 we present a Bioinformatics case study that showcases the usefulness of our approach, and finally in Section 4, some conclusions and future research lines are summarized.

2 Approach Overview

The overall approach we propose to assist the selection of web-services based on users requirements consists of three phases: (i) *Requirements elicitation and specification*, (ii) *Normalization* and (iii) *Web service selection* phases.

2.1 Requirements Elicitation and Specification

The user's information requirements are gathered and formally described in the *requirements model* using the i^* formalism [12]. Basically, this model regards the goals and tasks which are expressed in natural language and are completely independent of the characteristics of the web services. More details about this phase can be found in [7, 9].

2.2 Normalization

As task descriptions are expressed in natural language, they must be normalized in order to be automatically processed. The normalization consists of a semantic annotation process in which the description of the tasks are processed and annotated with concepts from a reference ontology to allow the reconciliation of the user's requirements with the web services descriptions. In contrast to [9], where tasks were only annotated with a BioCatalogue category, in this paper we improve the normalization by annotating the tasks with domain-specific concepts. Therefore, the semantic annotation process is divided in two steps:

1. **Domain specific annotations.** The purpose of this step is to identify the terms of the user-defined task related to the domain on which we are working, that is Life Sciences. We have used a semantic annotator which is capable of using several ontologies to annotate biomedical texts [2]. In our case, it annotates the task description with concepts from the UMLS(Unified Medical Language System) [1] and ^{my}Grid [11] ontologies. To determine the similarity between a text and a concept this annotator uses the following information-theoretic function:

$$sim(C,T) = max_{S \in lex(C)}(ratio(S,T))$$

$$ratio(S,T) = \frac{idf(cw(S,T)) - missing(S,T)}{idf(S)}$$

$$missing(S,T) = (idf(S) - idf(cw(S,T)))$$

The function $ratio(S,T)$ defines the ratio between the achieved information evidence for the text T, in our case the task description, and the information encoded in the lexicon form S. The function $missing(S,T)$ is the amount of information contained in S that have not been covered with T. $idf(S)$ measures the relevance of the terms in the string S, and $cw(S,T)$ is the set of terms in common between the concept string S and the text T. The function idf is defined as follows:

$$idf(S) = -\sum_{w \in S} log(P(w|UMLS))$$

The relevance of word w is measured by means of its estimated probability within the whole UMLS lexicon (i.e. $P(w|UMLS)$). After applying

[1] http://www.nlm.nih.gov/research/umls/

the semantic annotator, each task is represented with a semantic vector containing the $tf * idf$ values of each concept.

2. **Application specific annotations.** The next step is to determine the functionality the task is describing. For that purpose, we use the taxonomy of categories defined by BioCatalogue to classify the user-defined tasks. BioCatalogue has a taxonomy of web services categories related to Life Sciences, and many web services registered in BioCatalogue are annotated with at least one category. The matching between the tasks descriptions and the categories is performed by the ISub metric [10].

2.3 Web Service Selection

1. **Web services discovery.** The discovery of the suitable web services for the user's requirements is based on the matching between the annotations of the tasks and the metadata of the web services. However, most web service registries lack of metadata. To address this problem, we have automatically annotated the description (or the documentation in case there is no description available), the categories and the tags of 1729 services registered in BioCatalogue with the previously mentioned annotator. The annotations of each service are represented as a vector that contains the $tf * idf$ of each concept.

   The discovery process is made by two independent searches: (i) search for the web services that are annotated with the same category as the task, and (ii) search for the web services that have concepts in common with the user-defined task. The results of both searches are combined and they are scored based on the following linear combination:

$$score = C\text{-}score * w_1 + S\text{-}score * w_2$$

   where w_1 and w_2 are weights that depend on the relevance of each search and fulfill the condition, $w_1 + w_2 = 1$, *C-score* is the score calculated by the ISub metric, that is, the score of the string matching between the categories of the taxonomy and the task, and *S-score* is the cosine similarity between the vector of the task and the vector of the service.

2. **Web service assessment.** At the end of the discovery process, the user is provided with a set of ranked lists of web services with respect to the user-defined tasks. The user will then have to select which ones are the most appropriate for her requirements.

   The first problem to be solved is searching for services that can be considered equivalent, since many services provide such a broad functionality that they may be adequate for several requirements. For this purpose, an appropriate approach is to try to match sets of requirements to sets of candidate web services, in order to simplify the problem and provide guidance to the user. We use a biclustering method to perform the matching between requirements and services. [8] gives further information about this technique.

At the end of the process, the information obtained by the assessment is provided to the user. If the results are not those expected by the user, she can refine the process at the three phases of the guide.

3 Case Study

In this paper we want to present the results of our approach in a Bioinformatics case study. In contrast to previous works [8, 9], in which the discovery was only based on the category of the tasks, in this paper we have improved our approach by adding semantics. Here, we focus mainly on the results of the normalization and the web selection phases.

3.1 Requirements Elicitation and Specification

The requirements model, which consists of goals and user-defined tasks, is shown in Figure 1. More details about this phase are provided in [8].

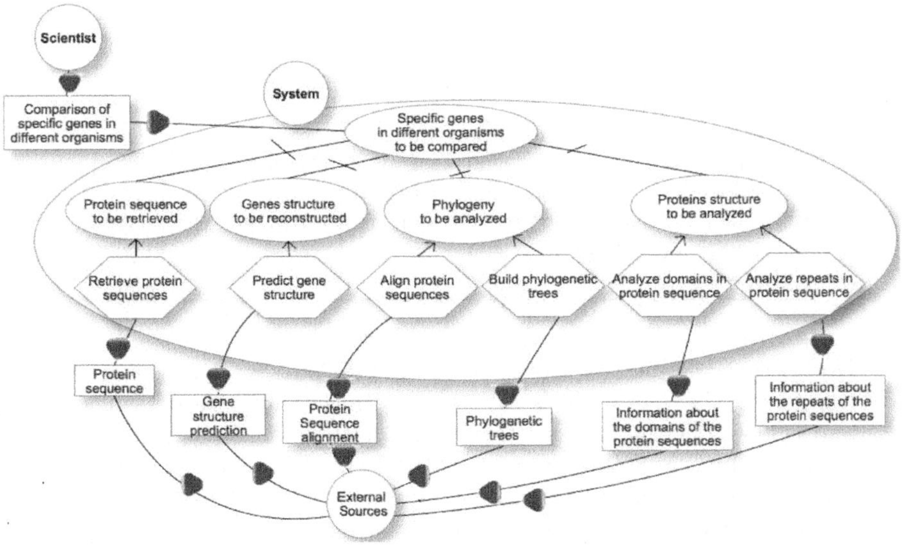

Fig. 1 Requirements model of the case study

3.2 Normalization

The normalization is carried out in two steps:

1. **Domain specific annotations.** The result of this phase consists of the semantic vector of each task. For example, the task "Align protein sequences" is represented with the semantic vector { *'D9000509':25,*

'D9000419':15}, where D9000509 is retrieved by "align sequences" with an $tf * idf$ value of 25 and, D9000419 is retrieved by "protein sequences" with an $tf * idf$ value of 15.

2. **Application specific annotations.** In this phase, the same tasks are aligned to categories of BioCatalogue. For example, the task "Align protein sequences" is annotated with the category "Protein Sequence Alignment" with a score of 0.7976.

3.3 Web Service Selection

Once the user's requirements have been gathered and normalized, the next step is the discovery and the assessment of the web services that are appropriate for these user's requirements.

1. **Web Service discovery.** The discovery is carried out by two independent searches in the annotated version of the BioCatalogue registry and, then, the results of both searches are combined. Table 1 shows the highest ranked services for the task "Align protein sequences" in which the scores of each search are specified.

Table 1 Ranked list of services for the task *Align protein sequences*

Service	Shared annotations	C-score	S-score	Score
infoAlignService	D9000509, D9000419	0.7976	0.65835	0.6862
BlastService	D9000509, D9000419	0.7976	0.5249	0.5795
BlastlocalgtService	D9000509, D9000419	0.7976	0.5249	0.5795
EMBOSS ehmmalign	D9000509	0.0	0.7078	0.5663
EMBOSS extraalign	D9000509	0.0	0.6843	0.5475

Notice that the highest scored service, "infoAlignService", has the same category as the task and it is annotated with the same concepts (D9000509 and D9000419), but it has a lower S-score than "EMBOSS ehmmalign", which is only annotated with D9000509, but it has more relevance than the concepts in "infoAlignService" metadata. Only the two first services are annotated with the same category as the task.

In this experiment, to calculate the final score we have used as weights for the relevance of the searches the values $w_1 = 0.2$ and $w_2 = 0.8$, as most of the web services are not categorised. Therefore, we give more relevance to the search based on the semantic annotations. At the end, a set of the ranked lists of services is obtained.

2. **Web service assessment.** In this phase, a biclustering technique is applied to find out the relationships between services and tasks and, in consequence, find out equivalent services. The biclustering technique has

as input a logical matrix that determines if two elements are related or not. In this case study, it has found 3 biclusters using a threshold of 0.0 (scores range from -1 to 1):

- Cluster 1: Search similar protein sequences, align protein sequences, analyze domains in protein sequences and analyze repeats in protein sequences.
- Cluster 2: Search similar protein sequences, analyze domains in protein sequences and analyze repeats in protein sequences.
- Cluster 3: Align protein sequences, analyze domains in protein sequences and analyze repeats in protein sequences.

For instance, cluster 1 includes the services: *EMBOSS fuzztran, EMBOSS patmatdb, EMBOSS fuzzpro, fuzztranService, INB.inb.bsc.es.runPratt, tBLASTn at SIB protein query sequences against a nucleotide sequence database dynamically translated in all reading frames, MRS Blast service, INB.inb.bsc.es.runPhylipProtpars Phobius, ELMdb, GlobPlot, EMBOSS tmap, GenesilicoProteinSilicoSOAPService, uniprot, TFmodeller.*

These clusters provide useful information about the correspondence between services and requirements. They determine that there are services that can provide the functionality required by several tasks, and therefore, the list of candidate services can be reduced. Finally, the assessment techniques will take as input the information provided by these clusters to produce a more specific assessment.

More details about the result of this case study can be found online[2].

4 Conclusions and Future Work

In this paper we have presented a semi-automatic approach to guide the user in the selection of the most appropriate web services to fulfill her information needs. The discovery process is based on the semantic normalization of both the requirements and the web services metadata. By annotating the tasks with domain-specific concepts, we have improved the representation of the tasks and enhanced the matching between the tasks and the web services metadata. Then, a biclustering technique is applied to find relations between tasks and web services. One of the main benefits of our approach is that it is an inherently exploratory process in which the user is able to change parameters and automatic selections. Moreover, although the current approach is applied to the Life Sciences domain, adapting it to new domains would only require the selection of adequate domain ontologies.

Currently, we are working on the validation of our approach, which implies the validation of the normalization and discovery phases. We are also working

[2] http://krono.act.uji.es/people/maria/semantic-discovery-of-web-services/case-study.pdf

on techniques that are not only based on general QoWS measures, but also on measures that analyze the quality of web services with respect to the requirements. These techniques introduce the specification of non-functional requirements, which destermine different criteria to assess the web services, since they describe quality attributes that are relevant to the user.

Acknowledgements. This research has been supported by the Spanish Ministry of Education and Science (TIN2008-01825/TIN) and by UJI – Fundació Bancaixa (P11B2008-43). María Pérez has been supported by UJI predoctoral grant PREDOC/2007/41.

References

1. Al-Masri, E., Mahmoud, Q.H.: QoS-based Discovery and Ranking of Web Services. In: ICCCN (2007)
2. Berlanga, R., Nebot, V., Jimenez, E.: Semantic annotation of biomedical texts through concept retrieval. In: BioSEPLN (2010)
3. Burgun, A., Bodenreider, O.: Accessing and integrating data and knowledge for biomedical research. Med. Inform., 91–101 (2008)
4. Cochrane, G.R., Galperin, M.Y.: The 2010 Nucleic Acids Research Database Issue and online Database Collection: a community of data resources. In: NAR 2010 (2010)
5. Lord, P., Alper, P., Wroe, C., Goble, C.A.: Feta: A light-weight architecture for user oriented semantic service discovery. In: Gómez-Pérez, A., Euzenat, J. (eds.) ESWC 2005. LNCS, vol. 3532, pp. 17–31. Springer, Heidelberg (2005)
6. Nucleic Acids Research.: Web Server Issue 38 suppl. 2 (2010)
7. Pérez, M., Casteleyn, S., Sanz, I., Aramburu, M.J.: Requirements gathering in a model-based approach for the design of multi-similarity systems. In: MoSE+DQS (2009)
8. Pérez, M., Sanz, I., Berlanga, R.: A biclustering-based technique for requirement-driven Web Service selection. In: JISBD, pp. 109–120 (2010)
9. Pérez, M., Sanz, I., Berlanga, R., Aramburu, M.J.: Semi-automatic discovery of web services driven by user requirements. In: Bringas, P.G., Hameurlain, A., Quirchmayr, G. (eds.) DEXA 2010. LNCS, vol. 6261, pp. 62–75. Springer, Heidelberg (2010)
10. Stoilos, G., Stamou, G.B., Kollias, S.D.: A string metric for ontology alignment. In: Gil, Y., Motta, E., Benjamins, V.R., Musen, M.A. (eds.) ISWC 2005. LNCS, vol. 3729, pp. 624–637. Springer, Heidelberg (2005)
11. Wolstencroft, K., Alper, P., Hull, D., Wroe, C., Lord, P.W., Stevens, R.D., Goble, C.A.: The myGrid ontology: bioinformatics service discovery. IJBRA (2007)
12. Yu, E.: Modelling Strategic Relationships for Process Reenginering. Ph.D. thesis, University of Toronto, Canada (1995)

Bringing Statistical Methodologies for Enterprise Integration of Conversational Agents

David Griol and José Manuel Molina

Abstract. In this paper we present a methodology to develop commercial conversational agents that avoids the effort of manually defining the dialog strategy for the dialog management module. Our corpus-based methodology is based on selecting the next system answer by means of a classification process in which the complete dialog history is considered. This way, system developers can employ standards like VoiceXML to simply define system prompts and the associated grammars to recognize the users responses to the prompt, and the statistical dialog model automatically selects the next system prompt. We have applied this methodology for the development of an academic conversational agent.

Keywords: Conversational Agents, Dialog Management, Statistical Methodologies, User Modeling.

1 Introduction

Conversational agents [11, 10] provide a appropriate interface between humans and computers, which is present in a growing number of commercial systems. In order to enable the rapid deployment of voice applications, markup languages such as VoiceXML have been widely adopted as they reduce the programming difficulty. However, the development of dialog systems with this approach involves a very costly engineering cycle [15]. As an attempt to reduce this cost and carry out rapid system prototyping, statistical approaches are gaining increasing interest.

There is a high variety of applications in which conversational agents can be used. One of the most wide-spread is information retrieval. Some sample applications are tourist and travel information [6], weather forecast over the phone [21], speech controlled telephone banking systems [12], conference help [1], etc. They have

David Griol · José Manuel Molina
Group of Applied Artificial Intelligence (GIAA), Computer Science Department,
Carlos III University of Madrid
e-mail: {david.griol,josemanuel.molina}@uc3m.es

J.M. Corchado et al. (Eds.): Trends in PAAMS, AISC 90, pp. 153–160.
springerlink.com

also been used for education and training, particularly in improving phonetic and linguistic skills: assistance and guidance to F18 aircraft personnel during maintenance tasks [2], dialog applications for computer-aided speech therapy with different language pathologies [17].

Spoken interaction can be the only way to access information in some cases, like for example when the screen is too small to display information (e.g. handheld devices) or when the eyes of the user are busy in other tasks (e.g. driving) [18]. It is also useful for remote control of devices and robots, specially in smart environments [13]. Finally, one of the most demanding applications for fully natural and understandable dialogs, are embodied conversational agents and companions [3, 5].

Learning statistical approaches to model the different modules that compose a dialog system has been of growing interest during the last decade [20]. Models of this kind have been widely used for speech recognition and also for language understanding [14, 9]. Even though in the literature there are models for dialog managers that are manually designed, over the last few years, approaches using statistical models to represent the behavior of the dialog manager have also been developed [16, 19]. These approaches are usually based on modeling the different processes probabilistically and learning the parameters of the different statistical models from a dialog corpus.

In this paper, we propose a statistical methodology to develop dialog managers for commercial conversational agents. Our methodology is mainly based on the modelization of the sequences of the system and user dialog acts and the introduction of a partition in the space of all the possible sequences of dialog acts. This partition, which is defined taking into account the data supplied by the user throughout the dialog, makes the estimation of a statistical model from the training data manageable. The confidence measures provided by the recognition and the understanding modules are also taken into account in the definition of this partition. The new system utterance is selected by means of a classification procedure. Specifically, we use neural networks for the implementation of this classification process. This methodology is combined with the standard VoiceXML to develop the rest of the modules of the conversational agent. We have applied our proposal to develop a conversational agent which provides academic information.

2 Our Proposal to Introduce Statistical Methodologies for the Development of Conversational Agents in Industry Applications

As stated in the introduction, our approach to integrate statistical methodologies in industry applications is based on the automatic learning of the dialog strategy using a statistical dialog management methodology. In most dialog systems, the dialog manager takes its decisions based only on the information provided by the user in the previous turns and its own model. This is the case with most slot-filling dialog systems. The methodology that we propose for the selection of the next system answer in this kind of tasks is detailed in [8]. Our technique is based on the definition of a data structure that we call Dialog Register (*DR*), and contains the information provided by the user throughout the previous history of the dialog. For each time i, the selection of the next system prompt A_i is carried out by means of the following maximization:

$$\hat{A}_i = \underset{A_i \in \mathscr{A}}{\operatorname{argmax}} P(A_i | DR_{i-1}, S_{i-1})$$

where set $\mathscr{A}$ contains all the possible system answers and S_{i-1} is the state of the dialog sequence (*system-turn, user-turn*) at time i.

Each user turn supplies the system with information about the task; that is, he/she asks for a specific concept and/or provides specific values for certain attributes. However, a user turn could also provide other kinds of information, such as task-independent information. This is the case of turns corresponding to *Affirmation*, *Negation* and *Not-Understood* dialog acts. This kind of information implies some decisions which are different from simply updating the DR_{i-1}. For that reason, for the selection of the best system answer A_i, we take into account the *DR* that results from turn 1 to turn $i-1$, and we explicitly consider the last state S_{i-1}. Our model can be extended by incorporating additional information to the *DR*, such as some chronological information (e.g. number of turns up to the current turn) or user profiles (e.g. nave or experimented users or user preferences).

The selection of the system answer is carried out through a classification process, for which a multilayer perceptron (MLP) is used. The input layer receives the codification of the pair (DR_{i-1}, S_{i-1}). The output generated by the MLP can be seen as the probability of selecting each of the different system answers defined for a specific task.

As stated before, the *DR* contains information about concepts and attributes provided by the user throughout the previous history of the dialog. For the dialog manager to determine the next answer, we have assumed that the exact values of the attributes are not significant. They are important for accessing the databases and for constructing the output sentences of the system. However, the only information necessary to predict the next action by the system is the presence or absence of concepts and attributes. Therefore, the information we used from the *DR* is a codification of this data in terms of three values, $\{0, 1, 2\}$, for each field in the *DR* according to the following criteria:

- **0:** The concept is unknown or the value of the attribute is not given.
- **1:** The concept or attribute is known with a confidence score that is higher than a given threshold. Confidence scores are given during the recognition and understanding processes. and can be increased by means of confirmation turns.
- **2:** The concept or attribute is activated with a confidence score that is lower than the given threshold.

As an attempt to improve the current technology, we propose to merge statistical approaches with VoiceXML. Our goal is to combine the flexibility of statistical dialog management with the facilities that VoiceXML offers, which would help to introduce statistical approaches for the development of commercial (and not strictly academic) conversational agents. To this end, our proposal employs the described statistical dialog management technique to decide the next system prompt. In addition, the system prompts and the grammars for ASR are implemented in VoiceXML compliant formats, for example, JSGF or SRGS.

In contrast with other hybrid approaches, our main aim is not to incorporate knowledge about best strategies in statistical dialog management, but rather to take advantage of an implementation language which has been traditionally used to build rule-based systems (such as VoiceXML), for the development of statistical dialog strategies. Expert knowledge about deployment of VoiceXML applications, development environments and tools can still be exploited using our technique. The only change is in the transition between states, which is carried out on a data-driven basis (i.e., is not deterministic). Figure 1 shows the complete architecture designed for the integration of the statistical methodology for dialog management with the functionalities provided by the VoiceXML standard.

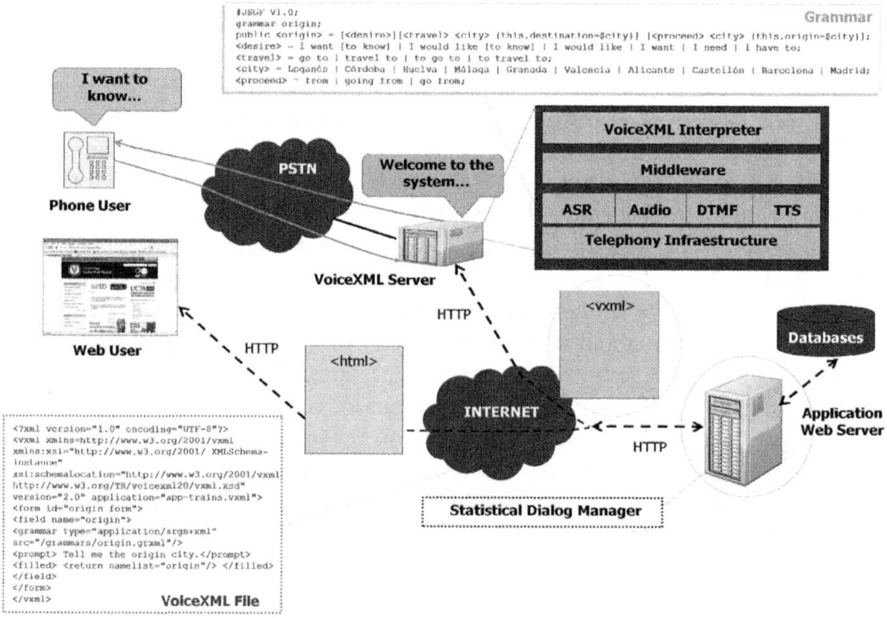

Fig. 1 Architecture for the integration of the proposed statistical methodology for dialog management with the VoiceXML standard

To learn the dialog model we use dialog simulation techniques. Our approach for acquiring a dialog corpus is based on the interaction of a user simulator and a dialog manager simulator [7]. Both modules use a random selection of one of the possible answers defined for the semantics of the task (user and system dialog acts). At the beginning of the simulation, the set of system answers is defined as equiprobable. When a successful dialog is simulated, the probabilities of the answers selected by the dialog manager during that dialog are incremented before beginning a new simulation.

The user simulation simulates the user intention level, that is, the simulator provides concepts and attributes that represent the intention of the user utterance. An error simulator module has been designed to perform error generation. In addition, the error simulator adds a confidence score to each concept and attribute in the frames.

A maximum number of turns per dialog is defined for acquiring a corpus using our user simulator, taking in to account the requirements of the task for real users. A user request for closing the dialog is selected once the system has provided the information defined in the objective(s) of the dialog. The dialogs that fulfill this condition before the maximum number of turns are considered successful. The dialog manager considers that the dialog is unsuccessful and decides to abort it when the following conditions take place: i) The dialog exceeds the maximum number of user turns; ii) The answer selected by the dialog manager corresponds with a query not required by the user simulator; iii) The database query module provides an error warning because the user simulator has not provided the mandatory information needed to carry out the query; iv) The answer generator provides a error warning when the selected answer involves the use of a data not provided by the user simulator.

3 Development of an Academic Conversational Agent

To test our proposal, we have used the UAH Task [4]. UAH is a spoken dialog system developed to provide spoken access to academic information about the Department of Languages and Computer Systems in the University of Granada. The information that the system provides can be classified in four main groups: subjects, professors, doctoral studies, and registration.

As in many other conversational agents, the semantic representation that we have chosen for the UAH task is based on the concept of frame. In the case of user turns, we defined four concepts related to the different queries that the user can perform to the system (*Subject*, *Lecturers*, *Doctoral studies*, *Registration*), three task-independent concepts (*Affirmation*, *Negation*, and *Not-Understood*), and eight attributes (*Subject-Name*, *Degree*, *Group-Name*, *Subject-Type*, *Lecturer-Name*, *Program-Name*, *Semester*, and *Deadline*).

The labeling of the system turns is similar to the labeling defined for the user turns. A total of 30 task-dependent concepts was defined: Task-independent concepts (*Affirmation*, *Negation*, *Not-Understood*, *New-Query*, *Opening*, and *Closing*); Concepts used to inform the user about the result of a specific query (*Subject*, *Lecturers*, *Doctoral-Studies*, and *Registration*); Concepts defined to require the user the attributes that are necessary for a specific query (*Subject-Name*, *Degree*, *Group-Name*, *Subject-Type*, *Lecturer-Name*, *Program-Name*, *Semester*, and *Deadline*);

Concepts used for the confirmation of concepts (*Confirmation-Subject*, *Confirmation-Lecturers*, *Confirmation-DoctoralStudies*, *Confirmation-Registration*) and attributes (*Confirmation-SubjectName*, *Confirmation-Degree*, *Confirmation-Group Name*, *Confirmation-SubjectType*, *Confirmation-LecturerName*, *Confirmation-ProgramName*, *Confirmation-Semester*, and *Confirmation-Deadline*).

The *DR* defined for the UAH task is a sequence of 12 fields, corresponding to the four concepts (*Subject*, *Lecturers*, *Doctoral-Studies*, and *Registration*) and eight attributes ((*Subject-Name*, *Degree*, *Group-Name*, *Subject-Type*, *Lecturer-Name*, *Program-Name*, *Semester*, and *Deadline*) defined for the task. The semantics selected for the dialog manager is represented through the 27 possible system answers defined for the UAH task. A set of 100,000 successfully simulated dialogs were used to learn the statistical dialog model.

A total of 25 dialogs were recorded from interactions of six students and professors of our University employing the conversational agent developed for the UAH task following our proposal. We considered the following measures for the evaluation:

1. Dialog success rate (%success). This is the percentage of successfully completed tasks. In each scenario, the user has to obtain one or several items of information, and the dialog success depends on whether the system provides correct data (according to the aims of the scenario) or incorrect data to the user.
2. Average number of turns per dialog (nT).
3. Confirmation rate (%confirm). It was computed as the ratio between the number of explicit confirmations turns (nCT) and the number of turns in the dialog (nCT/nT).
4. Average number of corrected errors per dialog (nCE). This is the average of errors detected and corrected by the dialog manager. We have considered only those errors that modify the values of the attributes and that could cause the failure of the dialog.
5. Average number of uncorrected errors per dialog (nNCE). This is the average of errors not corrected by the dialog manager. Again, only errors that modify the values of the attributes are considered.
6. Error correction rate (%ECR). The percentage of corrected errors, computed as nCE/ (nCE + nNCE).

The results presented in Table 1 show that the developed conversational can interact correctly with the users in most cases, achieving a success rate of 94%. The dialog success depends on whether the system provides the correct data for every objective defined in the scenario. The analysis of the main problems detected in the acquired dialogs shows that, in some cases, the system did not detect that the user wanted to finish the dialog. A second problem was related to the introduction of data in the DR with a high confidence value due to errors generated by the automatic speech recognizer that were not detected by the dialog manager. However, the evaluation confirms a good operation of the approach since the information is correctly given to the user in the majority of cases, as it is also shown in the value of the error correction rate.

Table 1 Results of the evaluation of the academic conversational agent

	%success	nT	%confirm	%ECR	nCE	nNCE
Conversational Agent	94%	11.6	38%	93%	0.89	0.06

4 Conclusions

In this paper, we have described a technique for integrating statistical techniques for the development of interactive conversational agents using a well known standard like VoiceXML, and considering a dialog model that is automatically learned from a dialog corpus. Our proposal works on the benefits of statistical methods for dialog management and VoiceXML, respectively. The former provide an efficient means to exploring a wider range of dialog strategies, whereas the latter makes it possible to benefit from the advantages of using the different tools and platforms that are already available to simplify system development.

We have applied our technique to develop a conversational agent that provides academic information, and have shown that it enables creating automatically VoiceXML documents to interact with the user. As a future work, we plan to study ways for adapting the proposed dialog management technique to more complex domains. Additionally, we are interested in investigating the main challenges that might arise in using our proposal to develop commercial systems, and the possible ways for easing the adoption of our technique in industry.

Acknowledgments. Funded by projects CICYT TIN2008-06742-C02-02/TSI, CICYT TEC 2008-06732-C02-02/TEC, CAM CONTEXTS (S2009/TIC-1485), and DPS2008-07029-C02-02.

References

1. Bohus, D., Grau, S., Huggins-Daines, D., Keri, V., Krishna, G., Kumar, R., Raux, A., Tomko, S.: Conquest - an Open-Source Dialog System for Conferences. In: Proc. of Human Language Technologies 2007: The Conference of the North American Chapter of the Association for Computational Linguistics, Rochester, NY, USA, pp. 9–12 (2007)
2. Bohus, D., Rudnicky, A.: Larri: A language-based maintenance and repair assistant. In: Proc. of IDS 2002, Kloster Irsee, Germany (2002)
3. Brahnam, S.: Building Character for Artificial Conversational Agents: Ethos, Ethics, Believability, and Credibility. PsychNology Journal 7(1), 9–47 (2009)
4. Callejas, Z., López-Cózar, R.: Relations between de-facto criteria in the evaluation of a spoken dialogue system. Speech Communication 50(89), 646–665 (2008)
5. Bailly, G., Raidt, S.: Gaze, conversational agents and face-to-face communication. Speech Communication 52(6), 598–612 (2010)
6. Glass, J., Flammia, G., Goodine, D., Phillips, M., Polifroni, J., Sakai, S., Seneff, S., Zue, V.: Multilingual spoken-language understanding in the MIT Voyager system. Speech Communication 17, 1–18 (1995)

7. Griol, D., Hurtado, L., Sanchis, E., Segarra, E.: Acquiring and Evaluating a Dialog Corpus through a Dialog Simulation Technique. In: Proc. of the 8th SIGdial Workshop on Discourse and Dialogue, Antwerp, Belgium, pp. 39–42 (2007)
8. Griol, D., Hurtado, L., Segarra, E., Sanchis, E.: A Statistical Approach to Spoken Dialog Systems Design and Evaluation. Speech Communication 50(8-9), 666–682 (2008)
9. He, Y., Young, S.: A data-driven spoken language understanding system. In: Proc. of ASRU 2003, pp. 583–588. St. Thomas, U.S. Virgin Islands (2003)
10. López-Cozar, R., Araki, M.: Spoken, Multilingual and Multimodal Dialogue Systems. In: Development and Assessment. John Wiley and Sons, Chichester (2005)
11. Mazuel, L., Sabouret, N.: Generic command interpretation algorithms for conversational agents. Web Intelligence and Agent Systems 6(1), 43–57 (2008)
12. Melin, H., Sandell, A., Ihse, M.: Ctt-bank: A speech controlled telephone banking system - an initial evaluation. In: TMH-QPSR, vol. 1, pp. 1–27 (2001)
13. Menezes, P., Lerasle, F., Dias, J., Germa, T.: Towards an interactive humanoid companion with visual tracking modalities. International Journal of Advanced Robotic Systems, 48–78 (2007)
14. Minker, W., Waibel, A., Mariani, J.: Stochastically-Based Semantic Analysis. Kluwer Academic Publishers, Dordrecht (1999)
15. Rojas-Barahona, L., Giorgino, T.: Adaptable dialog architecture and runtime engine (adarte): A framework for rapid prototyping of health dialog systems. International Journal of Medical Informatics 78, 56–68 (2009)
16. Torres, F., Hurtado, L., García, F., Sanchis, E., Segarra, E.: Error handling in a stochastic dialog system through confidence measures. Speech Communication 45, 211–229 (2005)
17. Vaquero, C., Saz, O., Lleida, E., Marcos, J., Canalís, C.: VOCALIZA: An application for computer-aided speech therapy in spanish language. In: Proc. of IV Jornadas en Tecnología del Habla, Zaragoza, Spain, pp. 321–326 (2006)
18. Weng, F., Varges, S., Raghunathan, B., Ratiu, F., Pon-Barry, H., Lathrop, B., Zhang, Q., Scheideck, T., Bratt, H., Xu, K., Purver, M., Mishra, R., Raya, M., Peters, S., Meng, Y., Cavedon, L., Shriberg, L.: CHAT: A Conversational Helper for Automotive Tasks. In: Proc. of the 9th International Conference on Spoken Language Processing (Interspeech-ICSLP), Pittsburgh, USA, pp. 1061–1064 (2006)
19. Williams, J., Young, S.: Partially Observable Markov Decision Processes for Spoken Dialog Systems. Computer Speech and Language 21(2), 393–422 (2007)
20. Young, S.: The Statistical Approach to the Design of Spoken Dialogue Systems. Tech. rep., Cambridge University Engineering Department (2002)
21. Zue, V., Seneff, S., Glass, J., Polifroni, J., Pao, C., Hazen, T., Hetherington, L.: JUPITER: A telephone-based conversational interface for weather information. IEEE Transactions on Speech and Audio Processing 8, 85–96 (2000)

Mining Web Pages Using Features of Rendering HTML Elements in the Web Browser

F.J. Fernández, José L. Álvarez, Pedro J. Abad, and Patricia Jiménez

Abstract. The Web is the largest repository of useful information available for human users, but it is usual that Web Pages do not provide an API to get access to its information automatically. In order to solve this problem, Information Extractors are developed. We present a new methodology to induce Information Extractors from the Web. It is based on rendering HTML elements in the Web browser. The methodology uses a KDD process to mining a dataset with features of the elements in the Web page. An experimentation over 10 web sites has been made and the results show the effectiveness of the methodology.

Keywords: Wrapper generation, web data extraction, data mining.

1 Introduction

The Web offers interest information for human users through Web Pages. However, in recent years, its growth has generated a large volume of information of interest, especially in business and, therefore the research community. The aim is developing automatic techniques to facilitate the processing of this large volume of information to extend the functionality of traditional web.

Information Extraction (IE) from Web Pages focus on extracting the relevant information from semi-structured Web Pages, unify them and offers it in a structured format to the end-users. The algorithms that perform the task of IE are referred to as extractors. Extractors can be classified in four classes [5] according to the intervention of an expert user: hand-crafted IE Systems, supervised IE Systems, semi-supervised IE Systems and unsupervised IE Systems.

Unsupervised IE Systems are based on the hypothesis that "Web Pages have a common template". Examples of these systems are MDR [2], IEPAD [13], RoadRunner

F.J. Fernández · José L. Álvarez · Pedro J. Abad · Patricia Jiménez
Departament of Information Technologies, University of Huelva
Ctra. Huelva-La Rábida, 21819 Palos de la Frontera, Huelva, Spain
e-mail: {javier.fernandez,alvarez,pedro.abad,
patricia.jimenez}@dti.uhu.es

J.M. Corchado et al. (Eds.): Trends in PAAMS, AISC 90, pp. 161–168.
springerlink.com

[14], or EXALG [1]. Supervised IE Systems need of annotating Web Pages by an expert user. Examples of these systems are WIEN [11], SoftMealy [4], STALKER [8], WHISK [12], SRV [7], DEPTA [18], and ViDE [15]. In Semi-supervised IE Systems, the annotation process can be automated somewhere: OLERA [3].

Therefore, in supervised IE Systems, the extractors must query the Web to collect the Pages, labeling the interest contents on the HTML code, and finally, training to learn patters that allow extracting content of interest from new Web Pages.

This paper presents a new supervised IE System based on rendering the HTML elements in the web browser. Three types of features have been defined: layout features, style features and content features. These features are calculated from HTML elements on the Web Pages. Knowledge Discovery in Databases (KDD) process is applied on this dataset to learn patters to extract interest content in new web Pages.

The rest of the paper is organized as follows: section 2 presents the methodology. Section 3 shows the cases of study and the results. Finally, section 4 concludes this work.

2 Methodology

Almost of the extractors are based on pattern recognition to extract the information from the structure of the web page. The goal of extractors is to extract the relevant information with the minimum error. However, today, web pages are full of non-relevant information trying to make the web page more striking. Moreover, the large number of advertisement in web pages makes difficult the identification of relevant information. In this situation, the extractors must to deal with two problems:

- First, the vast amount of non-relevant information to analyze. This produces that extractors spend a lot of time with the lack of the high computational cost of these algorithms. When the number of pages increases, the effectiveness loss becomes more evident.

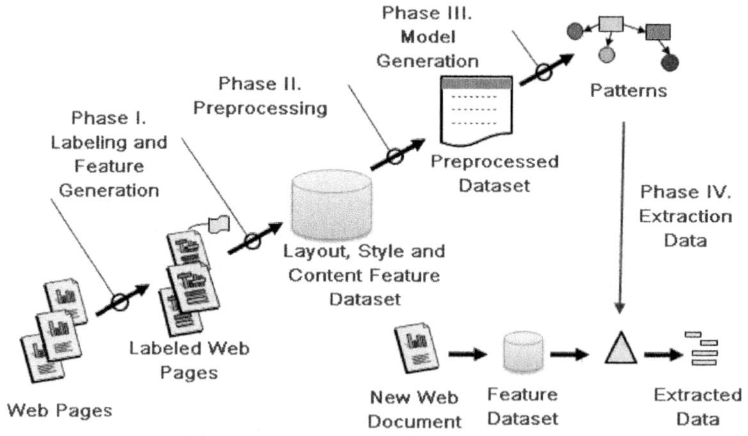

Fig. 1 An overview of the steps that compose the proposed methodology

- The second and harder problem is the loss of effectiveness in the extraction of the goal information. The non-relevant information in the pages makes the web page structure more complex and the extractors need to deal with these complexity.

The code of the non-relevant information in a Web page causes extraction algorithms fail when they try to extract the relevant information.

The proposed methodology is developed in the following steps (Fig. 1):

1. **Web Page labeling and feature generation.** This step involves the labeling of the information to extract and the feature generation of this information. The information selection is performed by the user. After labeling the features of selected information are generated and stored in a database. Features of both, relevant information and non-relevant information are generated and stored.
2. **Preprocessing of the training set.** Techniques of feature selection and classes balancing are applied to the database generated in step 1, which is considered as a training set in a supervised learning process.
3. **Models generation.** A model of the data contained in the training set is generated by a supervised learning algorithm. This model lets classify new information as belonging to one of the classes.
4. **Information Extraction.** The model generated in the previous step is applied to new web pages of the same Web site in order to extract the desired data.

To support this methodology we have developed the W-SOFIE (Web Software Observant For Information Extraction) tool. It automates the steps 1 to 3 and returns the induced model from the training pages.

The next sections describe in detail each one of the steps of the methodology.

2.1 Web Document Labeling and Feature Generation

The Web Pages of interest are those which present information about specific products. These pages display the product details as a product information sheet.

The W-SOFIE tool is able to navigate through these pages in order to label the desired elements. Labeling is performed by the user who also assigns the class to each selected element. Once the user has labeled all interesting elements, the calculation of the interesting area is performed. This process involves selecting the area on the page that contains the labeled elements. Only this area will be treated, the rest of the page is ignored because it does not contain relevant information. The coordinates of the interesting area of each page are stored.

The remaining not-selected elements within the interesting area are automatically labeled as not-interesting. All of these not-interesting elements are assigned to the same class, the not-interesting class.

Once all elements on the page are labeled, W-SOFIE automatically generates the features characterising them, and then, a tuple is stored in the database. This tuple represents an element and consists of the calculated features and the class assigned. Each element on the page is stored in the database, both the interesting elements and not-interesting ones.

Three types of features are generated and stored for the labeled elements:

- **Position features.** X and Y coordinates, high and wide of the element, the distance from the left-top corner of interesting area and the slope of this hypothetical line.
- **Style features.** Font color and style, background color and borders.
- **Content features.** Text density, digit density.

After this step, a database containing a tuple for each element labeled, and the coordinates of the interesting area of all treated pages is generated. The number of treated pages should be sufficient to represent all the possible layouts of the product pages.

2.2 Pre-processing of the Training Set

Previous database is ready for KDD process. This database is considered as a training set. However, before the underlying model can be learned, a preprocessing step is necessary. The database has two main problems:

- The database has high dimensionality. There are a lot of features that represent each element. In addition, much of that information is not relevant. In order to solve this problem, several feature selection algorithms can be used.
- The database is imbalanced. There are a lot of tuples in database belonging to not-interesting class, and few tuples belonging to interesting class. A balancing of database is required to solve this problem.

  After this step, database is suitable for an automatic learning process.

2.3 Models Generation

Because the only elements into interesting area have been considered, a learning of the coordinates of these areas is needed. This way, the first task to perform is determining the coordinates of this area. It is discovered from all stored areas of all labeled pages. An area wrapping all of them is considered as the general interesting area.

After discovering the general interesting area, a variety of classification algorithms can be used to extract a model of the training data. The representation of the model depends on the algorithm used.

It is important to highlight that the learning process is performed off-line, before the information extraction step. Thus, the time taken to generate the model is spending only once, before the information extraction process.

2.4 Information Extraction

The information extraction step starts automatically extracting all elements on a new page within the interesting area. The interesting area considered for a new page is the general interesting area calculated in the previous step. None of all extracted elements have a class assigned. In order to consider these elements as belonging to a class, they are evaluated using the learned model in the previous step. The assigned class by classifier indicates if an element is interesting or not. All desired elements are stored and the not-interesting ones are rejected.

3 Experimentation and Results

In order to test the methodology, it was applied to 10 e-commerce WEB sites related with sales of books. These 10 web sites are the top ten by the Alexa (http://www.alexa.com) ranking.

The 50 Bestsellers pages of each site were tested. The elements to extract were: Title, Authors, Saving, Rates and Availability.

In order to reduce the dimensionality of the training database, four feature selection algorithms were tested: InfoGainAttributeEval (wrapper algorithm), ReliefFAttributeEval (wrapper algorithm), CfsSubsetEval (filter algorithm) and ConsistencySubsetEval (filter algorithm).

The wrapper algorithms return a ranking of the features. Instead, the filter algorithms return a subset of features. In order to select a unique subset of features a hand-crafted selection was performed. The features selected were all of them returned by filter algorithms and the most relevant features (50% of the better rank) in the ranking of the wrapper algorithms. The mean of the size of the resulting database was 18.74% of the original one with a standard deviation of 0.01.

The balancing of the database was performed by oversampling; i.e., replicating examples of the minority class.

Once pre-processing had finished, four classification algorithms were tested: SMO [10], IBk [6], C4.5 [9] and a Stacking meta-model, where three individual models (SMO, IBk, and JRip [17]) were combined. The meta-model was trained from the individual models using the C4.5 classifier.

Table 1 Datasets before pre-processing

	Accuracy percentage				AUC			
Web Sites	C45	SMO	IBK	META	C45	SMO	IBK	META
Amazon	99.04	98.92	99.28	99.28	1.00	1.00	1.00	1.00
Barnesandnoble	98.68	98.52	98.68	97.87	1.00	1.00	1.00	0.94
Bestbuy	97.93	99.58	99.17	99.58	0.98	1.00	1.00	1.00
Buy	99.47	99.47	99.47	99.47	0.96	1.00	0.99	1.00
Ebay	98.18	91.64	98.82	99.14	0.97	0.91	1.00	0.99
Overstock	99.79	99,90	100.00	99.79	0.98	1.00	1.00	0.98
Play	99.71	99.71	99.71	99.41	1.00	1.00	1.00	1.00
Sears	99.57	99.57	100.00	98.22	1.00	1.00	1.00	1.00
Target	99.30	98.42	98.94	99.30	1.00	0.99	1.00	0.98
Walmart	99.66	99.20	99.43	99.31	1.00	1.00	1.00	1.00

Table 2 Datasets after feature selection

	Accuracy percentage				AUC			
Web Sites	C45	SMO	IBK	META	C45	SMO	IBK	META
Amazon	99.03	98.92	99.21	99.10	1.00	1.00	1.00	0.99
Barnesandnoble	98.65	98.57	98.55	98.37	1.00	1.00	1.00	0.99
Bestbuy	99.61	99.34	99.30	98.68	0.99	1.00	1.00	0.99
Buy	99.69	99.41	99.47	99.51	0.96	1.00	0.99	1.00
Ebay	98.97	90.47	99.05	99.11	0.98	0.90	1.00	0.99
Overstock	99.64	99.90	100.00	99.89	0.99	1.00	1.00	1.00
Play	99.70	99.70	99.70	99.53	1.00	1.00	1.00	1.00
Sears	99.56	99.48	100.00	98.99	1.00	1.00	1.00	1.00
Target	99.74	92.99	98.08	99.39	1.00	0.93	0.99	0.99
Walmart	99.73	99.20	99.43	99.47	0.99	1.00	1.00	1.00

Table 3 Datasets after feature selection and balancing

	Accuracy percentage				AUC			
Web Sites	J48	SMO	IBK	META	J48	SMO	IBK	META
Amazon	99.96	98.86	99.94	100.00	1.00	1.00	1.00	1.00
Barnesandnoble	99.86	98.44	99.72	99.89	1.00	1.00	1.00	1.00
Bestbuy	99.81	99.75	99.81	99.69	1.00	1.00	1.00	1.00
Buy	99.52	98.78	99.54	99.54	1.00	0.99	1.00	1.00
Ebay	99.94	97.82	99.88	99.93	1.00	1.00	1.00	1.00
Overstock	100.00	99.98	100.00	99.98	1.00	1.00	1.00	1.00
Play	99.89	99.89	99.89	99.85	1.00	1.00	1.00	1.00
Sears	99.84	99.73	100.00	99.32	1.00	1.00	1.00	1.00
Target	99.85	99.32	99.71	99.87	1.00	1.00	1.00	1.00
Walmart	99.99	99.86	99.91	99.96	1.00	1.00	1.00	1.00

The results obtained by 10-folds cross-validation are shown in tables 1 to 3. Values of Accuracy and Area Under ROC are detailed.

Values on tables show a very good accuracy and AUC of all algorithms and web sites. Most values are near 100% accuracy and 1 AUC.

Table 4 presents the mean of accuracy percent obtained for all classification algorithms in each web site.

A statistical analysis was carried out considering the mean of accuracy percent by means of Wilkoxon test [16]. The test did not detect significance ($p=0.05$) difference between results of dataset before preprocessing and results of dataset with feature selection (result of the test: $p=0.492$). This way, the conclusion that feature selection does not improve the classification is probed. Moreover, significance (p=0.05) difference was found between results of Datasets after feature selection and balancing and the others (result of the test: p=0.00586 in both cases). Therefore, we can conclude that feature selection and balancing presents the best accuracy.

Table 4 Mean of accuracy percent for each web site

Web Sites	Datasets before pre-processing.	Datasets after feature selection.	Datasets after feature selection and balancing.
Amazon	99.13	99.07	99.69
Barnesandnoble	98.44	98.54	99.48
Bestbuy	99.07	99.23	99.77
Buy	99.47	99.52	99.35
Ebay	96.95	96.90	99.39
Overstock	99.87	99.86	99.99
Play	99.64	99.66	99.88
Sears	99.34	99.51	99.72
Target	98.99	97.55	99.69
Walmart	99.40	99.46	99.93

4 Conclusions

In this paper we have presented a new supervised IE System to extract data in Web Pages based on rendering HTML elements in the Web browser. Layout, Style and Content features are calculated on the elements generating a dataset. A model is induced applying a KDD process to the dataset. Thus, data from new Web Pages can be extracted using this model.

Experimentation with 10 website of e-commerce has been made. The results show the effectiveness of this system.

The bottleneck of our method is the annotation phase, since the user intervention is required. In future works, a semi-automatic process will be used.

Acknowledgments. This work was partially sponsored by Spanish Ministry of Science and Technology and Junta de Andalucía under grants TIN2007-64119, P07-TIC-02602, and P08-TIC-4100.

References

1. Arasu, A., Garcia-Molina, H.: Extracting structured data from web pages. In: ACM SIGMOD International Conference on Management of Data, pp. 337–348 (2003)
2. Liu, B., Grossman, R., Zhai, Y.: Mining web pages for data records. IEEE Intelligent Systems 19, 49–55 (2004)
3. Chang, C.H., Kuo, S.C.: OLERA: Semisupervised web-data extraction with visual support. IEEE Intelligent Systems 19(6), 56–64 (2004)

4. Hsu, C.N., Dung, M.T.: Generating finite-state transducers for semi-structured data extraction from the web. Information Systems 23(8), 521–538 (1998)
5. Chang, C.-H., Kayed, M., Girgis, M.R., Shaalan, K.F.: A Survey of Web Information Extraction Systems. IEEE Transactions on Knowledge and Data Engineering 18, 1411–1428 (2006)
6. Aha, D.W., Kibler, D., Albert, M.K.: Instance based learning algorithms. Machine Learning 6(1), 37–66 (1991)
7. Freitag, D.: Information extraction from HTML: Application of a general learning approach. In: Fifteenth Conference on Artificial Intelligence, AAAI 1998 (1998)
8. Muslea, I., Minton, S., Knoblock, C.A.: A hierarchical approach to wrap-per induction. In: Agents, pp. 190–197 (1999)
9. Ross Quinlan, J.: C4.5: Programs for machine learning. Morgan Kaufmann Publishers Inc., San Francisco (1993)
10. Platt, J.C.: Fast Training of Support Vector Machines using Sequential Minimal Optimization, pp. 185–208 (1999)
11. Kushmerick, N., Weld, D.S., Doorenbos, R.B.: Wrapper induction for information extraction. In: The 15th International Joint Conference on Artificial Intelligence (IJCAI), pp. 729–737 (1997)
12. Soderland, S.: Learning information extraction rules for semi-structured and free text. Journal of Machine Learning 34(1-3), 233–272 (1999)
13. Csie, T.C., Hui Chang, C.: IEPAD: Information extraction based on pattern discovery. In: The 10th International Conference on World Wide Web, pp. 681–688 (2001)
14. Crescenzi, V., Mecca, G., Merialdo, P.: RoadRunner: Towards automatic data extraction from large web sites. In: The 27th International Conference on Very Large Data Base, pp. 109–118 (2001)
15. Liu, W., Meng, X., Meng, W.: VIDE: A Vision-Based Approach for Deep Web Data Extraction. IEEE Transactions on Knowledge and Data Engineering 22, 447–460 (2009)
16. Wilcoxon, F.: Individual comparisons by ranking methods. Biometrics 1, 80–83 (1945)
17. Cohen, W.W.: Fast Effective Rule Induction. In: Twelfth International Conference on Machine Learning, pp. 115–123 (1995)
18. Zhai, Y.: Web data extraction based on partial tree alignment. In: Proceedings of the 14th International Conference on World Wide Web (WWW 2005), pp. 76–85 (2005)

An Experiment on Using Datamining Techniques to Extract Information from the Web

Gretel Fernández and Hassan A. Sleiman

Abstract. The Web has become an indispensable source of knowledge due to the large amount of information available on it. Unfortunately, accessing this information by automated processes is not a simple task. Besides the semantic web and Web services, information extraction is a solution for this problem. Our proposal is to explore how datamining techniques to infer classifiers can help us build classifiers to extract information from web pages. In this paper we present a first approach to our technique that consisted on evaluating how these techniques work on DOM tree features. Our first results show a good precision and recall which indicates a new promising technique.

Keywords: Information extraction from the Web, Datamining techniques.

1 Introduction

The Web can be considered as the largest data repository ever created. Its data is usually embedded in presentation tags (HTML) in order to provide a human friendly interface that allows humans to interact, understand and navigate easily through the Web. The problem is that implementing software agents to process data that is in human friendly formats is not easy at all, which hinders using web data into automated business processes.

The importance of web information aroused enterprises' interests in integrating this information into their business processes to enrich them. An example can be the use of this information in their business intelligence and market assessment processes. Automating the integration of web information into business processes is a requirement in order to obtain web data automatically and which is usually performed using web wrappers [5].

Gretel Fernández · Hassan A. Sleiman
University of Sevilla
e-mail: {gretel,hassansleiman}@us.es

J.M. Corchado et al. (Eds.): Trends in PAAMS, AISC 90, pp. 169–176.
springerlink.com

A web wrapper aims at offering an API to abstract programmers from simulating human behavior interacting with the Web. One of the main components of a web wrapper is the information extractor. Information extractors are general-purpose algorithms that rely on a number of extraction rules. These rules can be hand-crafted but there are a number of proposals in the literature that report on learners to infer them. There are also a few proposals that do not rely on rules such as Stavies [18], IESS [15] and ViPER [19], and proposals such as S-CREAM [9], KIM [13], Armadillo [4] and MnM [22] that focus on free text, but they fall out of the scope of this paper. Rules may range from regular expressions to context few grammars or XPath queries.

To the best of our knowledge, no-one has ever explored using datamining techniques to infer classifiers in this field. In this paper, we report on an experiment in which we have used such techniques to build a classifier that can determine what the contents of a DOM node are, e.g, a title, an author's name, a price, or irrelevant information. The conclusion is that using these techniques is promising.

A common feature between a high number of information extractors is that they all work with the HTML code with the DOM Tree or with an image of the web page. None of them makes uses style sheets (CSS). We propose an idea that is based on CSS features of DOM tree nodes and that can be extracted from web pages to identify and classify items using datamining techniques. The first step in our proposal is the definition of a set of classes that represent the data we are interested in extracting from a web site, e.g.title, author, price or NA (irrelevant information). Then, a set of web pages from this web site are annotated by assigning for each DOM node a class. Once classes are assigned, we calculate a table with the normalised CSS features, text length, level and class of every DOM node. The number of NAs in the table is much greater than the number of the instances of other classes, so a random selection algorithm was applied over the NA instances and a percentage of them was chosen to create a CSV file from the previous table. The percentage we have used for our experiments is ten percent more than the number of nodes that were not classified as NA. Each CSV file is considered an independent dataset on which some of the algorithms that Weka [24] provides were used to build several classification models that must be empirically.

The rest of the paper is organised as follows: In Section 2 we review briefly related work and then in Section 3 shows our experiments and the first experimental results. Finally, Section 4 presents the conclusions and our future work regarding this novel technique.

2 Related Work

The literature provides many techniques to infer extraction rules from both semi-structured and free text web pages. They can be broadly classified into unsupervised and supervised techniques. Unsupervised techniques rely on datasets that need not be annotated by user. Unsupervised techniques for semi-structured web pages include ViDE [16], Alvarez et. al's [1], RoadRunner [6], ExAlg [2], FivaTech [12] and DeLa [23].

ViDE uses some visual and nonvisual features to infer a set of extraction rules first by separating data records and then aligning them using an algorithm similar to string alignment algorithms. Alvarez et. al's uses XPaths and a clustering technique to detect data regions, i.e. where the information of interest in a web page reside, and to separate them into data records. Then, it uses multiple string matching algorithm to extract attributes from these records.

RoadRunner infers extraction rules by searching for the template that was used at the server side to generate the similar input web pages. It compares these web pages and produces a regular expression in case a difference between these pages is found. The result is a regular expression that acts as a template; in short, it includes verbatim text that represents parts that are constant across pages and variables that represent changes from page to page. ExAlg tries also to infer the template from a set of similar web pages by distinguishing the template tokens from data tokens in input web pages. It classifies tokens using equivalence classes and occurrence vectors to detect the tokens that belong to the template.

FivaTech infers extraction rules by obtaining a fixed/variant pattern tree which is a structure that allows identifying the template used to generate the input web pages and also by detecting the data schema used in these pages in order to extract objects and their attributes. It uses clustering, string alignment and a pattern detection to infer this template. DeLa is another unsupervised algorithm that uses suffix trees to detect repeating patterns and then constructs a pattern tree which is a structure where these repeating patterns are saved. String alignment is then used to detect optional tokens.

Supervised techniques rely on datasets that must be annotated by the user. Supervised techniques for semi-structured web pages include Softmealy [11], OLERA [3], Whisk [20], DEByE [14], Threasher [10], SRV [7] and Stalker [17].

Softmealy uses user annotations to build an automaton according to the order in which annotations appear and then infers the transitions' rules aligning tokens and then generalising them. OLERA needs a single annotation to infer extraction rules. It uses approximate pattern matching and generalisations by searching for strings that are similar to the annotated example, then asks the user to confirm the results, and then it generates a regular expression to extract data records by using multiple string alignment.

Whisk uses a coverage algorithm to infer extraction rules. It is a top-down algorithm that starts from the most general rule, then, terms are added one by one to the rule until the covering errors are reduced to a predefined threshold. DEByE is an approach where user creates a structure that represents the annotated data and maps this data into this structure for extraction rules inference. It infers regular expressions to each annotated attribute by searching for the largest common prefix and suffix that covers all annotated examples. The structure is then used to rebuild extracted objects.

Threasher uses the subtrees annotated by the user to induce a pattern tree that identifies data records. It uses a tree edit distance to map nodes between different subtrees. SRV infers first order logic rules, it starts with all annotated attributes and an empty rule, and then adds predicates greedily trying to cover the maximum number of annotated examples and the minimum non-annotated examples at each step. Stalker relies on a tree-like structure called embedded catalog where each node contains an automaton to extract an attribute. The embedded catalog allows reconstructing the objects and attributes extracted by the automaton.

Previous proposals focus on inferring extraction rules for semi-structured web pages such as web pages where query results are shown or detailed information of a product is presented. Other proposals focus on inferring extraction rules for web pages with free texts such as news web pages and blogs. These techniques include S-CREAM [9], KIM [13], Armadillo [4] and MnM [22].

We shall not provide additional details on free text extraction techniques because we are only interested on inferring rules to extract information from semi-structured web pages. It is a technique that builds on extraction rules that are classification models built by means of a classic datamining technique. Note that this differs from the previous techniques in that it uses data mining techniques.

3 The Experiments

We conducted a series of experiments to evaluate our idea. We used thirty-two systhetic datasets and fifteen real datasets. The following algorithms to infer classification models for our datasets were used: Naive Bayes, C4.5, IBk, ZeroR, VFI, SVM [24]. In each experiment, we measured precision (P), recall (R) and time to build and test the model (TT). The average and standard deviation values are presented in table 1. We compared the results obtained from each dataset and made a ranking. The analysis of our experiments rely on statistical inference. The idea is to define two hypothesis, the null hypothesis which always considers that there are not significant differences between values and the alternative hypothesis that considers that there are differences.

The goal is to reject the null hypothesis and show that there are differences. Accepting or rejecting the null hypothesis depends on the comparison of the p-value calculated by the tests and significance level (α). In our study we used the standard value (α =0.05). If p-value is less than α then the null hypothesis is rejected. The first step was to determine if P, R and TT behaved normally using both the Kolmorogov-Smirnov's test with the Lilliefors correction and the Shapiro-Wilk's test. The null hypothesis in this case is that there are no differences between the P, R and TT samples and the theoretical corresponding normal distribution in which the parameters were estimated from our samples. Table 2 shows the results of the normality test. Note that there are several samples in which both p-values are 0.00. The implication is that we need to resort to non-parametric tests, i.e., we need to use the Kruskal-Wallis test to determine if P, R and TT means can be considered equal (null hypothesis) or not (alternative hypothesis). Table 3 show the p-values grouping by algorithm for each measure (precision, recall and execution time). These values demonstrate that the techniques are different. The next step is then to use Friedman's algorithm to calculate the scores and Bergmann's algorithm to rank them. The first test (Table 4) confirms the results obtained by the Kruskal-Wallis test and shows the values obtained are in order. Finally Bergmann's test compare the algorithms and returns a list of different techniques. With these results we obtain the ranking of the algorithms for each variable: precision, recall and time to build and test the model (see table 5).

Table 1 Descriptive Statistics

Technique	P	R	TT
N. Bayes	0.71±0.16	0.73±0.12	0.05±0.01
C4.5	0.74±0.19	0.75±0.17	0.06±0.01
IBk	0.78±0.16	0.79±0.16	0.04±0.01
ZeroR	0.35±0.04	0.59±0.03	0.03±0.01
VFI	0.77±0.15	0.72±0.16	0.04±0.01
SVM	0.59±0.14	0.72±0.10	0.09±0.03

Table 2 Kolmorogov-Smirnov's and Shapiro-Wilk's tests

	Kolmogorov-Smirnov (P-Value)			Shapiro-Wilk (P-Value)		
Technique	P	R	TT	P	R	TT
N. Bayes	0.06	0.20	0.00	0.09	0.30	0.00
C4.5	0.16	0.07	0.00	0.06	0.03	0.00
IBk	0.01	0.01	0.00	0.00	0.00	0.00
ZeroR	0.07	0.08	0.00	0.00	0.01	0.00
VFI	0.16	0.20	0.00	0.01	0.11	0.00
SVM	0.03	0.03	0.00	0.02	0.11	0.00

Table 3 Kruskal-Wallis test

	P	R	TT
P-Value	0.00	0.00	0.00

Table 4 Friedman's scores

P (p-value: 0.00)		R (p-value: 0.00)		TT (p-value: 0.00)	
Algorithms	Scores	Algorithms	Scores	Algorithms	Scores
IBk	2.50	IBk	2.63	ZeroR	0.18
VFI	2.56	C4.5	3.00	VFI	0.21
C4.5	2.69	N. Bayes	3.34	IBk	0.23
N. Bayes	2.96	SVM	3.37	N. Bayes	0.34
SVM	4.34	VFI	3.56	C4.5	0.44
ZeroR	5.96	ZeroR	5.10	SVM	1.00

Table 5 Bergmann's ranking

Ranking	P	R	TT
1	IBk, VFI, C4.5, N. Bayes	IBk, VFI, SVM, C4.5, N. Bayes	ZeroR
2	SVM	ZeroR	VFI, IBk
3	ZeroR		N. Bayes, C4.5
4			SVM

4 Conclusions and Future Work

After comparing the results we can conclude that the best algorithms with respect to precision and recall were the IBk and VFI but they had the second lowest time, while the algorithm ZeroR had the best time but was the worst regarding precision and recall. Furthermore Naive Bayes and C4.5 were as good in precision and recall as the first one but they were slower and finally the SVM algorithm was the last one with respect to time. These results prove that our proposal for extracting information is not perfect but it is promising and should be studied more deeply.

Building a set of datasets was the most complicated task since there is not a set of standard datasets for comparison. Besides, the models inferred by the classifiers are not applicable over other datasets because the values are not uniform e.g., for google chrome navigator the color can be rgb(0,0,0) while for the internet explorer navigator the color can be #000000F, so it is necessary to normalise all features values as a solution to this problem. Another problem at this moment is that we can extract relevant elements from web pages, e.g., title, author, price, but we do not know which book these elements belong to. The results showed that some of the features used to infer classifiers had no effect over the inferred model, which means that these features can be discarded in future work. The values of precision and recall used in our experiment were calculated globally, it would be interesting to work with the values of each class and compare their results. The high number of NA instances is another problem that we faced and that was solved using a random instance selection algorithm.

Our future work will focus on resolving previous problems, as well as obtaining other node features. These features can be tree nodes' features which may improve the classification. For example, the number of immediate children nodes, the parent node, sibling nodes and their CSS features, the height of a node (which is the length of the longest downward path to a leaf from that node) and the number of nodes descendent. Another proposal to consider for the NA instances selection algorithm is [21]. Once the technique is totally defined, the number of datasets will be increased and our technique will be compared with other techniques over these datasets.

Acknowledgements. Supported by the European Commission (FEDER), the Spanish and the Andalusian R&D&I programmes (grants grants TIN2010-21744-C02-01, TIN2007-64119, P07-TIC-2602, P08-TIC-4100, TIN2008-04718-E, and TIN2010-09988-E). The work by Gretel Fernández was supported by a grant awarded by the International Relationship Office of the University of Seville. We would like to thank Dr. José C. Riquelme for helping us define the methodology to carry out our experimental analysis.

References

1. Álvarez, M., Pan, A., Raposo, J., Bellas, F., Cacheda, F.: Extracting lists of data records from semi-structured web pages. Data Knowl. Eng. 64(2), 491–509 (2008), http://dx.doi.org/10.1016/j.datak.2007.10.002
2. Arasu, A., Garcia-Molina, H.: Extracting structured data from web pages. In: SIGMOD Conference, pp. 337–348 (2003), http://doi.acm.org/10.1145/872757.872799
3. Chang, C.H., Kuo, S.C.: Olera: Semisupervised web-data extraction with visual support. IEEE Intelligent Systems 19(6), 56–64 (2004), http://dx.doi.org/10.1109/MIS.2004.71
4. Ciravegna, F., Chapman, S., Dingli, A., Wilks, Y.: Learning to harvest information for the semantic web. In: Bussler, C.J., Davies, J., Fensel, D., Studer, R. (eds.) ESWS 2004. LNCS, vol. 3053, pp. 312–326. Springer, Heidelberg (2004)
5. Corchuelo, R., Arjona, J.L., Ruiz, D.: Wrapping web data islands. J. UCS 14(11), 1808–1810 (2008)
6. Crescenzi, V., Mecca, G., Merialdo, P.: Roadrunner: Towards automatic data extraction from large web sites. In: VLDB, pp. 109–118 (2001), http://www.vldb.org/conf/2001/P109.pdf
7. Freitag, D.: Information extraction from HTML: Application of a general machine learning approach. In: AAAI/IAAI, pp. 517–523 (1998)
8. Gómez-Pérez, A., Benjamins, V.R. (eds.): Proceedings of Knowledge Engineering and Knowledge Management Ontologies and the Semantic Web, 13th International Conference, EKAW 2002, Siguenza, Spain, October 1-4 (2002)
9. Handschuh, S., Staab, S., Ciravegna, F.: S-CREAM - semi-automatic CREAtion of metadata. In: Gómez-Pérez, Benjamins (eds.) [8], pp. 358–372, http://link.springer.de/link/service/series/0558/bibs/2473/24730358.htm

10. Hogue, A., Karger, D.R.: Thresher: Automating the unwrapping of semantic content from the World Wide Web. In: WWW, pp. 86–95 (2005), http://doi.acm.org/10.1145/1060745.1060762
11. Hsu, C.N., Dung, M.T.: Generating finite-state transducers for semi-structured data extraction from the Web. Inf. Syst. 23(8), 521–538 (1998), http://dx.doi.org/10.1016/S0306-43799800027-1
12. Kayed, M., Chang, C.H.: Fivatech: Page-level web data extraction from template pages. IEEE Transactions on Knowledge and Data Engineering 22, 249–263 (2010),doi http://doi.ieeecomputersociety.org/10.1109/TKDE.2009.82
13. Kiryakov, A., Popov, B., Terziev, I., Manov, D., Ognyanoff, D.: Semantic annotation, indexing, and retrieval. J. Web Sem. 2(1), 49–79 (2004), http://dx.doi.org/10.1016/j.websem.2004.07.005
14. Laender, A.H.F., Ribeiro-Neto, B.A., da Silva, A.S.: DEByE - data extraction by example. Data Knowl. Eng. 40(2), 121–154 (2002), http://dx.doi.org/10.1016/S0169-023X0100047-7
15. Li, W., Dong, Y., Wang, R., Tian, H.: Information extraction from semi-structured WEB page based on DOM tree and its application in scientific literature statistical analysis system. In: Anonymous (ed.) SSME (2009), http://doi.ieeecomputersociety.org/10.1109/SSME.2009.59
16. Liu, W., Meng, X., Meng, W.: Vide: A vision-based approach for deep web data extraction. IEEE Trans. Knowl. Data Eng. (2010), http://doi.ieeecomputersociety.org/10.1109/TKDE.2009.109
17. Muslea, I., Minton, S., Knoblock, C.A.: Hierarchical wrapper induction for semistructured information sources. Autonomous Agents and Multi-Agent Systems 4(1/2), 93–114 (2001)
18. Papadakis, N., Skoutas, D., Raftopoulos, K., Varvarigou, T.A.: Stavies: A system for information extraction from unknown web data sources through automatic web wrapper generation using clustering techniques. IEEE Trans. Knowl. Data Eng. 17(12), 1638–1652 (2005), http://doi.ieeecomputersociety.org/10.1109/TKDE.2005.203
19. Simon, K., Lausen, G.: Viper: augmenting automatic information extraction with visual perceptions. In: CIKM, pp. 381–388 (2005), http://doi.acm.org/10.1145/1099554.1099672
20. Soderland, S.: Learning information extraction rules for semi-structured and free text. Machine Learning 34(1-3), 233–272 (1999)
21. Vallejo, C.G., Troyano, J.A., Ortega, F.J.: InstanceRank: Bringing order to datasets. Pattern Recogn. Lett. 31, 133–142 (2010), http://portal.acm.org/citation.cfm?id=1663654.1663889
22. Vargas-Vera, M., Motta, E., Domingue, J., Lanzoni, M., Stutt, A., Ciravegna, F.: MnM: Ontology driven semi-automatic and automatic support for semantic markup. In: Gómez-Pérez, Benjamins (eds.) [8], pp. 379–391, http://link.springer.de/link/service/series/0558/bibs/2473/24730379.htm
23. Wang, J., Lochovsky, F.H.: Data extraction and label assignment for web databases. In: WWW, pp. 187–196 (2003), http://doi.acm.org/10.1145/775152.775179
24. Witten, I.H., Frank, E.: Data Mining: Practical Machine Learning Tools and Techniques, 2nd edn. Morgan Kaufmann, San Francisco (2005), http://www.cs.waikato.ac.nz/~ml/weka/book.html

BRENHET2, A MAS to Facilitate the Reutilization of LOs through Federated Search

F. De la Prieta, A. Gil, S. Rodríguez, and B. Martín

Abstract. Nowadays, the reutilization of learning resources makes possible a reduction of the time of creation educational contents. In this sense, one of the most advances is the use of federated search systems in order to retrieve different resources from a set of repositories more quickly. But nowadays, the main problem is the high heterogeneity in this context. Multiagent Systems witch consider organizational could be the appropriate answer to solve this problem.

Keywords: Multi-agent systems, Virtual Organization, Learning objects, Repositories, federated search.

1 Introduction

One of the most widespread trends in the application of information technologies on the education sector bases its efforts on the fragmentation of educational content in independent and modular resources. This resources or fragments can be used in different environments and contexts, facilitating the creation of different learning experiences. In this sense, the reutilization makes possible a significantly reduction of the time of creation, not only of educational resources, but also their aggregation in order to create complete courses. This kind of design of reusable contents is one of the most significant advances in the field of distance education in what is known as a paradigm of learning objects (LO) 2.

The search and aggregation of these small fragments is the cause of educational developments in the generation process of educational content. For this is necessary that LO contains some external descriptors, which allow its identification, search and retrieval. This description is achieved through metadata in order to ensure interoperability between systems. The metadata of LO provide information for different purposes: educational, administrative and structural.

F. De la Prieta · A. Gil · S. Rodríguez · B. Martín
Departamento Informática y Automática, Universidad de Salamanca, Salamanca, Spain
e-mail: {fer,abg,srg,eureka}@usal.es

J.M. Corchado et al. (Eds.): Trends in PAAMS, AISC 90, pp. 177–184.
springerlink.com

Currently, there are different standards, the most remarkable is Dublin Core 3 and very especially Learning Object Metadata, LOM 2.

In order to facilitate the reuse, LOs are usually stored in Learning Object Repositories (LORs). This kind of systems provides the key infrastructure to store, manage, locate and retrieve all types of digital resources. There are several mechanisms of storage and infrastructure deployment 6, which implies that to ensure interoperability between LORs it is necessary to include abstraction layers through standardized protocols. Notably among these protocols the OAI-MPH 8, IMS DRI 7 and Simple Query Interface (SQI) 4.

On one hand, the specification of standards and protocols assumes a high heterogeneity of repositories: different formats of educational resources, metadata standards, technologies of LORs, as well as different practices of use by authors. On the other hand, there are not quality criteria which allow the validity verification of the descriptors created for a LO, or the information that provided about him.

One of the biggest challenges in creating a system that allows the reuse of LOs is, as already mentioned, the wide heterogeneity in this context. This brings out new techniques and tools to facilitate interoperability between systems. In this sense, research focusing on MAS from an organizational perspective seems to be gaining most ground, and the heterogeneous environments that concern us, the most suitable for them. Given the advantages provided by the unique characteristics found in the development of MAS from an organizational perspective and the lacks in the field of e-learning, this paper proposes a MAS model that facilitates the reutilization of LOs through federated search in different LORs.

The next section describes in more detail the organizational aspects of agents; section 3 presents an in-depth study of the educational context (LOR and LO); section 4 shows the proposed architecture; and finally section 5 contains some results and conclusions obtained.

2 Virtual Organizations of Agents

One of the challenges which are facing up a system that allows search and retrieval of LO is the high heterogeneity. This makes it necessary to use techniques which allow the adaptation of system in open and heterogeneous contexts like this.

The use of multi agent systems (MAS) is appropriate, due to they can respond and act by itself in face of any changes on its environment. Recently, MAS that consider organizational aspects are getting popularity 59. The main advantage of the development of MAS from social point of view, is it allow the ability to create heterogeneous systems in regards of languages, applications and features.

The Virtual Organization (VO) is defined as a set of individuals and institutions which need to coordinate their resources and services within institutional limits 1. The VO can be considered as open systems formed by grouping and collaboration of heterogeneous entities where there is a clear separation between structure and functionality 1.

The VOs are formed by social entities; these social entities in turn can be composed by a specific number of members or agents. These entities are not independent, they interact with each other following a structure. The topology followed by this structure defines how agents are sorted and are communicated with each other, or with external agents. Among the features of VO, the dynamism is crucial for its use on heterogeneous environment like this. This feature is related to the entry and exit of agents, as well as the adoption of roles by themselves, the creation of groups and the control of their behavior.

Making usage of the possibilities which are provided by the particularities of the organizational Multi-Agent System development, a model has been formalized for the optimal search and retrieval of educational resources in heterogeneous environments.

3 Study Background

Due to the high heterogeneity, it was decided to carry out an empirical study about the different LOR and LO, in order to know the real situation of e-learning systems.

The study was divided into three phases: (i) the first phase involved the analysis of LOR, both individual way and as a whole; (ii) the second phase consisted of a performance test, for it, 60 consultations were made with each of the LOR, choosing as search patterns a set of the UNESCO codes on field of science and technology; (iii) and finally in the last phase, with the recovered LO in the previous phase, the metadata schema that describe them was analyzed.

The study was focused on the kinds of LOR that implement the interoperability layer SQI 4, which is highly widespread. The following LOR was been included: Acknowledge, adds, Ariadne, CGIAR, EducaNext, LACLO-FLOR, LORNET, MACE, Merlot, Nime, OER Commons and Edna Online.

Although this study has been focused on LOR that implement SQI layer, the main deficiency is associated with those repositories that until today do not implement any abstraction layer that allows encapsulate the internal logic of the repository. Such repositories are monolithic and do not allow the automation of the query processes. However, others systems that include the abstraction layer, also suffer various problems (reliability, availability, access times, etc.) which were been observed during this study. However, although these LOR have different problems, in neither case the interoperability between platforms was prevented, which facilitates the reutilization of resources to the end user. The Figure 1.A shows the performance observed in some of the studied repositories. The performance is a measure corresponding to divide the query time in milliseconds between the number of result obtained. In view of these results we can deduce its poor performance, due to is much minor than 1.

To conclude the study of the LOR, it was found that some LOR where for each query made there are new federated searches in others LOR at the same time. This fact although it would be logical thinking that could improve the results, the reality is that increases the query time and causes duplicate results.

On the other hand, there are some related problems with LO, IEEE LOM, IEEE LOM 2 defines an educational resource perfectly, but do not define what are the mandatory attributes that must be specified to have a minimum quality.

From this, two fundamental problems were arising: (i) the univocal identification, as the fact that it is not obligatory to fill the designated field for this purpose; (ii) missing the location element, which prevents the reutilization of LO. The Figure 1.B shows the percentage of use of each of the defined categories in the IEEE LOM 2. In other words, Figure 1.B shows the percentage of use each of metadata category by the authors.

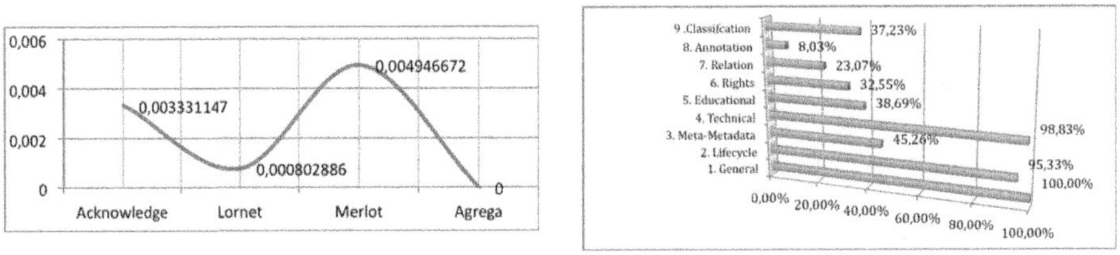

Fig. 1 A. Repositories performance ; B. Metadata Categories

All these problems that had been observed in both the LOR and the LO are an actual real challenge for applications which trying to use this kind of platforms. The reason is that these applications need being designed in order to be used in a highly heterogeneous environment. In this sense, designing the system´s architecture under the framework which providing the MAS is suitable, because we can apply theirs features (autonomy, location, reactivity, rationality, intelligence, organization, mobility and learning) to develop a stable system with the ability to react intelligently to the needs of the environment.

4 System Proposed: BRENHET^2

4.1 Architecture Details

For the development of this federated search system, it was decided the use of MAS that take into account norms and social organizations. This type of systems describes the functionality through a set of roles, so that each of the roles can be achieved by any agent, depending on the capabilities required. In practice, this new concept of design is crucial, as changes that occur in the context can also be introduced into the system easily.

The proposed system, BREHENT2, is seen as an intermediary communication point between the LOR, the LO that store, and users who make use of them. The system provides a federated search system that incorporates a phase of cataloging or management of results.

In a social environment, like proposed in this study, in order to establish the model of interaction between members of the system, it is necessary to analyze the needs and expectations of potential users of the system. From this analysis we have identified the roles of the users involved in the system:

- **User.** Represents the system's user. It has the functions of starting the search process, consulting the results and also the access to the statistics.
- **Query Manager.** It is responsible of supervise the entire search process. It has the functions of retrieve the query from the user, start the search process, request the sort of results and retrieve statistics for the query.
- **Repository Manager.** It has specific control over the process of federated search, it receives the query from the Query Manager agent and it is responsible to ask the previously selected LOR agents to initiate the search. It also has to check that these agents are functioning correctly to ensure performance.
- **LOR.** It is in charge of making the request to each repository. It implements the code of the middleware layers. There will be many LOR agents as repositories where the query is requested. It also has to request the transformation of query to Translator agent. It collects statistics and finally LOR also has to cancel the query in case of the repository does not work properly.
- **Translator.** Is the agent responsible for transforming the query to the language preferred by the concerned repository where the query is directed. There are different instances of agents, one for each language.
- **Results.** It receives LO from each of the LOR agents. It automatically extracts information from the metadata schema and removes invalid of them, finally it calculates the reutilization index which we will discuss below.
- **Cataloguer**. Is the agent in charge of cataloging the results.
- **Statistics.** This agent is responsible for collecting statistical.
- **Supervisor.** It has the overall control over the system. It supervises the operations of other agents in the system and checks the messages.

Following the model of agents that takes into account the organizational aspects, one of the first tasks will be to determine the mission of the organization model, which represents the products and services offered by the system. The Figure 3 shows the functional view that includes both the mission of the organization, and the functionalities necessary to meet the objectives.

This OV has the missions of maximizing the performance of the search and the maximizing the quality of the results. In order to satisfy these requirements (OOffers in the diagram) have the following services: search LO, review the results and consult statistics. Depending on the services offered by the system, the dynamism of the Virtual organization is modeled as an open system.

Starting from the functionality of the system we can determine what functionality can be implemented by internal agents and what must be provided by an external agent. Thus, the dynamism of the VO has been modeled as an open system. The roles LOR, User, translator and results are accessible by external agents, requiring a process of acquiring the role. The remaining roles (Query Manager, Supervisor, Repository Manager, cataloguer and Statistics) are associated with non-accessible internal agents.

Finally, it is necessary to determine the structural model that the system follows, since each of the components of the architecture can be structured in a different way. In this case, it was decided to use a holarchy type structure that is best suited to this system, since the members of each of the components is organized hierarchically.

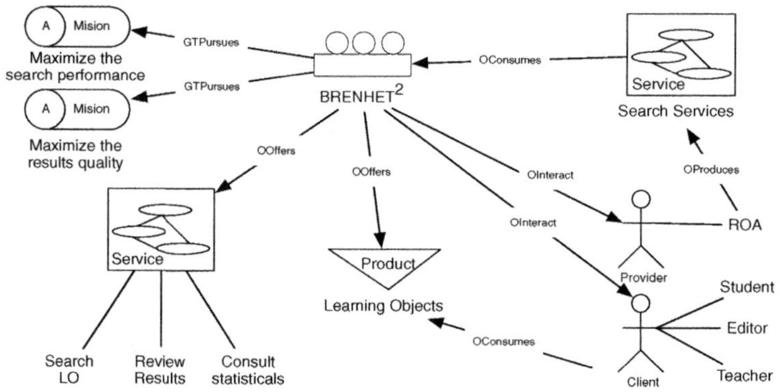

Fig. 2 Diagram of the organizational model (functionally)

4.2 Sort System Details

After obtaining the LO of the federated search process, it is necessary to filter and rank them in order to reduce the information overload to the end user of the application. Remark that this documentation process is performed for the IEEE LOM schema 2, but could be adapted to any other scheme.

The cataloging process consists on a process of filtering and assessment of reutilization of each LO. It then goes describing this process, which is performed entirely by the agent who takes the Results' role. Dividing the process into three steps:

1. The first step is the recovery of the information contained in the metadata that describe the LO.
2. The second step consists of verifying the quality of the extracted information. It is checked if the requirements defined in the standard are achieved, for each of the descriptors of LO, if these requirements are not achieved, they are not valid for reutilization and they will be eliminated. Also any LO will be eliminated in case of it has not got the location field that contains a link to the described educational resource.
3. The last step consists on obtaining a reutilization index of each LO of the results. For this will need to identify the most important elements of IEEE LOM 2 schema in order to facilitate the reutilization which is a complex process where general aspects, educational, administrative and technical have been taken into account.

Once the reference scheme has been obtained, then we need to determine the relationship degree between the proposed scheme and the scheme of each LO. For this we develop a weight vector so that each of its elements corresponds to the importance of each descriptor within the specific schema of LO.

To determine the relationship between these two vectors we use the Pearson correlation coefficient that indicates the degree of relationship between two random variables (weight vectors). This index is within the interval [-1, 1], the closer it is to 1 implies that the two vectors are more closely related and therefore the reutilization rate will be higher.

Finally, as a result a set of LO had been obtained, with a minimum quality metadata and an index that denotes the capacity to be reused.

5 Conclusions and Experimental Results

In this paper a new approach has been presented in order to design a federated search tool of LOR which it based on the implementation of the OV. At a theoretical level this new model of MAS fits perfectly into this heterogeneous context. The aim was to show that it also adapted at practical level.

The case study was created as a federated search system in which there are 2 LOR agents, an user agent, a type translator agent and results agent. The two LOR agents implement the client code for access to SQI repositories 4. Meanwhile, the translator agent implements the query language VSQL 10. To validate the system we have performed the same set of queries made in the study section of this work.

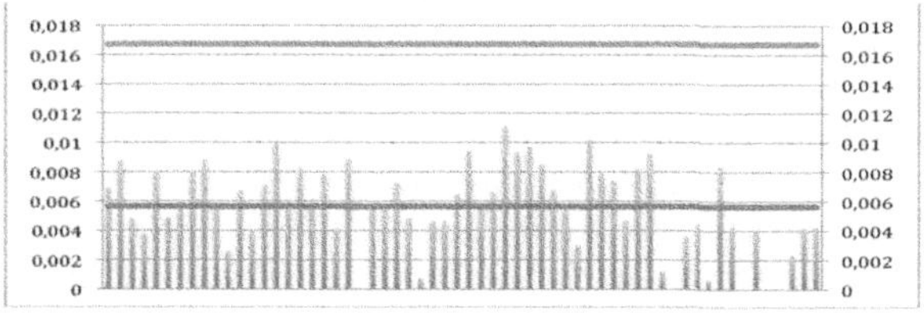

Fig. 3 Case study results (Performance)

The results of the tests have been fully satisfactory, being that the number of results has been significantly improved without increasing the query time. In summary way, the performance graph in the Figure 4 shows clearly that while the performance of each individual repository is around 0.000803 and 0.00494. The average of developed system is about 0.00567939 and its maximum is 0.011093183.

The research work carried out provides some new contributions in the fields of the Theory of Agents and Virtual Organizations (i) a methodology aimed at developing the VO, which has helped to cover all the needs of such system, had been studied; (ii) a case study had been defined in order to test the validity of such systems; (iii) the results obtained had been analyzed in order to demonstrate that the architecture together with the system are a viable alternative.

As conclusion, all the intended objectives had been achieved: The adaptation of the VO in search and retrieval of educational resources has been demonstrated and this has allowed verifying by empirical data that it is valid and that the performance has been increased.

Acknowledgments. This work has been partially supported by the project MICINN TIN 2009-13839- C03-03.

References

1. Boella, G., Hulstijn, J., Van Der Torre, L.: Virtual organizations as normative multiagent systems. In: HICSS. IEEE Computer Society, Los Alamitos (2005)
2. Draft Standard for Learning Object Metadata (IEEE 1484.12.1-2002). The Institute of Electrical and Electronics Engineers, Inc.
3. Dublin Core Metadata Initiative. DCMI Metadata Terms
4. European Committee for standardization – Cen Workshop Agregament. A simple Query Interface Specification for Learning Repositories. Ref. No.: CWA 15454:2005 E (2005)
5. Ferber, J., Gutknecht, O.: A meta-model for the analysis and design of organizations in multi-agent systems. In: Proc. of the Third International Conference on Multi-Agent Systems, pp. 128–135. IEEE Computer Society, Los Alamitos (1998)
6. Frango, I., Omar, N.: Architecture of Learning Objects Repositories. Learning Objects: standards, metadata, repositories & LMS, 131–155 (2007)
7. Riley, K., McKell, M., Mason, J.: IMS Digital Repositories Interoperability. Core Functions Information Model. Version 1.0 Final Specification (2003)
8. Lagoze, C., Van de Sompel, H., Nelson, M., Warner, S.: Open Archives Initiative. The Open Archives Initiative Protocol for Metadata Harvesting (2002)
9. Rodríguez, S., Pérez-Lancho, B., De Paz, J.F., Bajo, J., Corchado, J.M.: Ovamah: Multiagent- based Adaptive Virtual Organizations. In: 12th International Conference on Information Fusion, Seattle, Washington, USA (July 2009)
10. Simon, B., Massart, D., Van Assche, F., Ternier, S., Duval, E.: Simple Query Interface Specification. Version 1.0 Beta (2005)

Author Index

Alonso Berrocal, José L. 117
Álvarez, José L. 161
Álvarez, Laura 17

Abad, Pedro J. 161
Adam, Emmanuel 91
Adiego, Joaquín 109
Alonso, Ricardo S. 1
Alonso, Vidal 17
Aramburu, María José 145

Bajo, Javier 17
Bellas, Fernando 137
Berger, Thierry 91
Berlanga, Rafael 145

Chodura, Dawid 83
Cvetković, Božidara 33

Díaz, Raquel Gómez 117
De la Prieta, F. 177
de Luis, Ana 51
Demazeau, Yves 41
De Paz, Juan F. 51
Dominik, Paweł 83
Dossou, Paul-Eric 75
Dovgan, Erik 33

Fdez-Riverola, Florentino 9
Felgueiras, Miguel 9
Fernández, F.J. 161
Fernández, Gretel 169
Fernández-García, Antonio Jesús 127
Figuerola, Carlos G. 117

Gams, Matjaž 33
García, Óscar 1
Gil, A. 177
Golinska, Paulina 99
Griol, David 153

Hajdul, Marcin 99
Hallenborg, Kasper 41
Hamberg, Roelof 59
Hossain, Shabbir 41
Hoyos-Torío, Javier E. 109

Iribarne, Luis 127

Jiménez, Patricia 161

Kaluža, Boštjan 33
Koźlak, Jarosław 83

López, Javier 137
Lindgren, Helena 25
Luštrek, Mitja 33

Martín, B. 177
Martín, Patricia 17
Martínez-Prieto, Miguel A. 109
Maturana, Carmelo 127
Mirchevska, Violeta 33
Mitchell, Philip 75
Molina, José Manuel 153
Montoto, Paula 137

Nilsson, Ingeborg 25

Pérez, María 145
Pach, Cyrille 91
Pan, Alberto 137
Pawlewski, Pawel 67, 75
Pedrero, Alberto 17
Pereira, António 9

Rodríguez, S. 177

Sánchez, Miguel 1, 17
Sánchez-Martínez, Felipe 109
Sancho, David 1
Sanz, Ismael 145
Seco, Alexandra 9
Sleiman, Hassan A. 169
Surie, Dipak 25

Tapia, Dante I. 1
Trentesaux, Damien 91

van Heusden, Eef 59
van Wijngaarden, Bruno 59
Verriet, Jacques 59

Zambrano, Gabriel 91
Zato, Carolina 17, 51
Zazo Rodríguez, Angel F. 117

GPSR Compliance

The European Union's (EU) General Product Safety Regulation (GPSR) is a set of rules that requires consumer products to be safe and our obligations to ensure this.

If you have any concerns about our products, you can contact us on ProductSafety@springernature.com

In case Publisher is established outside the EU, the EU authorized representative is:

Springer Nature Customer Service Center GmbH
Europaplatz 3
69115 Heidelberg, Germany

Batch number: 10755047

Printed by Printforce, the Netherlands